Learn Autodesk Inventor 2018 Basics

3D Modeling, 2D Graphics, and Assembly Design

T. Kishore

Apress®

Learn Autodesk Inventor 2018 Basics

T. Kishore
Hyderabad, India

ISBN-13 (pbk): 978-1-4842-3224-8   ISBN-13 (electronic): 978-1-4842-3225-5
https://doi.org/10.1007/978-1-4842-3225-5

Library of Congress Control Number: 2017958952

Managing Director: Welmoed Spahr
Editorial Director: Todd Green
Acquisitions Editor: Steve Anglin
Development Editor: Matthew Moodie
Technical Reviewer: Wallace Jackson
Coordinating Editor: Mark Powers
Copy Editor: Kim Wimpsett

Distributed to the book trade worldwide by Springer Science+Business Media New York, 233 Spring Street, 6th Floor, New York, NY 10013. Phone 1-800-SPRINGER, fax (201) 348-4505, e-mail orders-ny@springer-sbm.com, or visit www.springeronline.com. Apress Media, LLC is a California LLC and the sole member (owner) is Springer Science + Business Media Finance Inc (SSBM Finance Inc). SSBM Finance Inc is a **Delaware** corporation.

For information on translations, please e-mail rights@apress.com, or visit www.apress.com/rights-permissions.

Apress titles may be purchased in bulk for academic, corporate, or promotional use. eBook versions and licenses are also available for most titles. For more information, reference our Print and eBook Bulk Sales web page at www.apress.com/bulk-sales.

Any source code or other supplementary material referenced by the author in this book is available to readers on GitHub via the book's product page, located at www.apress.com/9781484232248. For more detailed information, please visit www.apress.com/source-code.

Printed on acid-free paper

Contents

About the Author

T. Kishore is an experienced trainer, savvy engineer, and prolific author of several books on Autodesk and other tools for engineering, design, graphics, 3D printing, and more.

About the Technical Reviewer

Wallace Jackson has been writing for leading multimedia publications about his work in new media content development since the advent of Multimedia Producer Magazine nearly two decades ago. He has authored a half-dozen Android books for Apress, including four books in the popular Pro Android series. Wallace received his undergraduate degree in business economics from the University of California at Los Angeles and a graduate degree in MIS design and implementation from the University of Southern California. He is currently the CEO of Mind Taffy Design, a new media content production and digital campaign design and development agency.

Introduction

Autodesk Inventor as a topic of learning is vast, with a wide scope. It is a package of many modules that deliver great value to enterprises. It offers a set of easy-to-use tools for designing, documenting, and simulating 3D models. Using this software, you can speed up the design process and reduce your product development costs.

This book provides a step-by-step approach for users to learn Autodesk Inventor. It is aimed at those with no previous experience with Inventor. However, users of previous versions of Inventor may find this book useful to learn about the new and enhanced features of Inventor 2018. You will be guided from starting an Autodesk Inventor 2018 session to creating parts, assemblies, and drawings. Each chapter explains the components with the help of real-world models.

Scope of This Book

This book was written for students and engineers who are interested in using Autodesk Inventor 2018 to design mechanical components and assemblies and then create drawings.

This book provides a step-by-step approach for learning Autodesk Inventor 2018. The chapters cover the following topics:

- **Chapter 1** introduces Autodesk Inventor. The user interface and terminology are discussed in this chapter.

- **Chapter 2** takes you through the creation of your first Inventor model. You create simple parts.

- **Chapter 3** teaches you to create assemblies. It explains the top-down and bottom-up approaches for designing an assembly. You create an assembly using the bottom-up approach.

- **Chapter 4** teaches you to create drawings of the models created in the earlier chapters. You also place exploded views and the part list of an assembly.

- **Chapter 5** teaches you how to use additional modeling tools to create complex models.

- **Chapter 6** introduces you to sheet metal modeling. You create a sheet metal part using the tools available in the Sheet Metal environment.

- **Chapter 7** teaches you to create top-down assemblies. It also introduces you to creating mechanisms by applying joints between the parts.

- **Chapter 8** teaches you to apply dimensions and annotations to a 2D drawing.

- **Chapter 9** teaches you to add 3D annotations and tolerances to a 3D model.

CHAPTER 1

Getting Started with Autodesk Inventor 2018

This chapter covers the most commonly used features of Autodesk Inventor.
In this chapter, you will do the following:

- Understand the Inventor terminology

- Start a new file

- Understand the user interface

- Understand different environments in Inventor

In this chapter, you will learn some of the most commonly used features of Autodesk Inventor. In addition, you will learn about the user interface.

In Autodesk Inventor, you create 3D parts and use them to create 2D drawings and 3D assemblies. **Inventor is feature-based**. Features are shapes that are combined to build a part. You can modify these shapes individually.

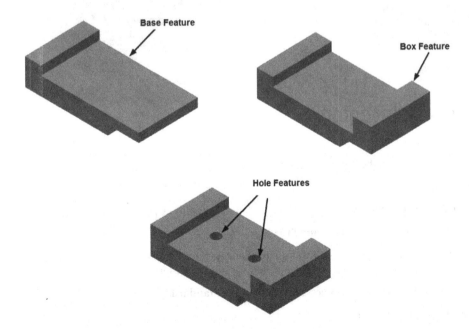

© T. Kishore 2017
T. Kishore, *Learn Autodesk Inventor 2018 Basics*, https://doi.org/10.1007/978-1-4842-3225-5_1

Most of the features are sketch-based. A sketch is a 2D profile and can be extruded, revolved, or swept along a path to create features.

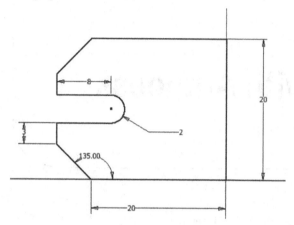 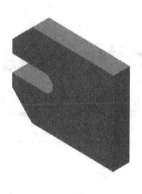

Inventor is parametric in nature. You can specify standard parameters between the elements. Changing these parameters changes the size and shape of the part. For example, see the following design of the body of a flange before and after modifying the parameters of its features:

 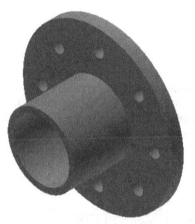

Starting Autodesk Inventor

To start Autodesk Inventor, follow these steps:

1. Click the **Start** button on the Windows taskbar.

2. Click **All Programs**.

3. Click **Autodesk ➤ Autodesk Inventor 2018 ➤ Autodesk Inventor 2018**.

4. On the ribbon, **click Get Started ➤ Launch ➤ New**.

5. In the **Create New File** dialog, click the **Templates** folder located at the top-left corner. You can also select the **Metric** folder to view various metric templates.

6. In the **Part – Create 2D and 3D Objects** section, click the **Standard.ipt** icon.

7. Click **Create** to start a new part file.

Notice these important features of the Inventor window:

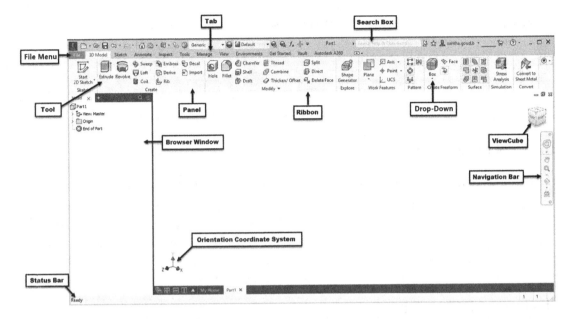

Exploring the User Interface

Various components of the user interface are discussed next.

Ribbon

The ribbon is located at the top of the window. It consists of various tabs. When you click a tab, several tools appear. These tools are arranged in panels. You can select the required tool from a panel. The following sections explain the various tabs of the ribbon available in Autodesk Inventor.

The Get Started Ribbon Tab

This ribbon tab contains tools such as **New, Open, Projects**, and so on.

The 3D Model Ribbon Tab

This ribbon tab contains the tools to create 3D features, planes, surfaces, and so on.

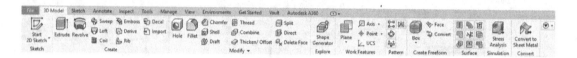

The View Ribbon Tab

This ribbon tab contains the tools to modify the display of the model and user interface.

The Inspect Ribbon Tab

This ribbon tab has tools to measure the objects. It also has analysis tools to analyze the draft, curvature, surface, and so on.

The Sketch Ribbon Tab

This ribbon tab contains all the sketch tools.

The Assemble Ribbon Tab

This ribbon tab contains the tools to create an assembly. It is available in an assembly file.

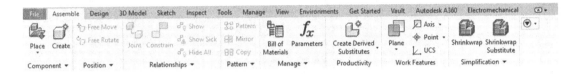

The Presentation Ribbon Tab

This tab contains the tools to create the exploded views of an assembly. It also has the tools to create presentations, assembly instructions, and animations of an assembly.

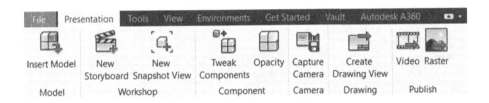

The Drawing Environment Ribbon Tab

In the Drawing environment, you can create print-ready drawings of a 3D model. The ribbon tabs in this environment contain tools to create 2D drawings.

The Place Views Ribbon Tab

This ribbon tab has commands and options to create and modify drawing views on the drawing sheet.

The Annotate Ribbon Tab

This ribbon tab has commands and options to add dimensions and annotations to the drawing views.

The Sheet Metal Ribbon Tab

The tools in this tab are used to create sheet metal components.

File Menu

The File menu appears when you click the **File** tab located at the top-left corner. This menu contains the options to open, print, export, manage, save, and close a file.

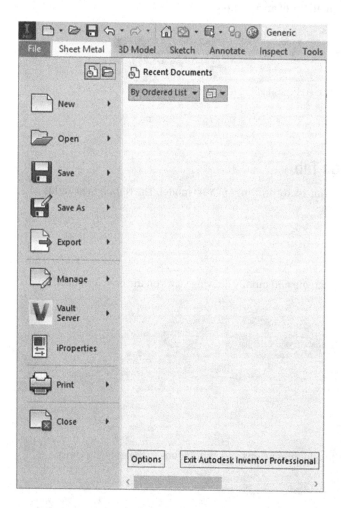

Quick Access Toolbar

This is available at the top left of the window. It contains the tools such as **New, Save, Open**, and so on.

You can customize this toolbar by clicking the down arrow on the right side of this toolbar.

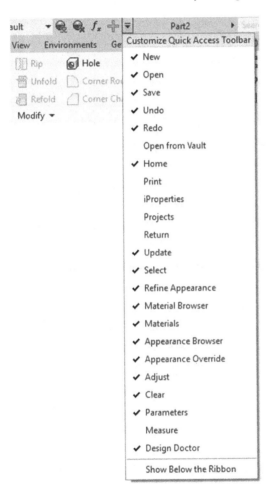

Browser Window

The Browser window is located on the left side of the window. It contains the list of operations carried in an Autodesk Inventor file.

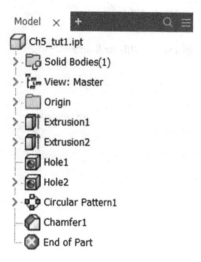

Status Bar

This is available below the Browser window. It displays the prompts and actions taken while using the tools.

Navigation Bar

This is located at the right side of the window. It contains the tools to zoom, rotate, pan, or look at a face of the model.

ViewCube

The ViewCube is located at the top-right corner of the graphics window. It is used to set the view orientation of the model.

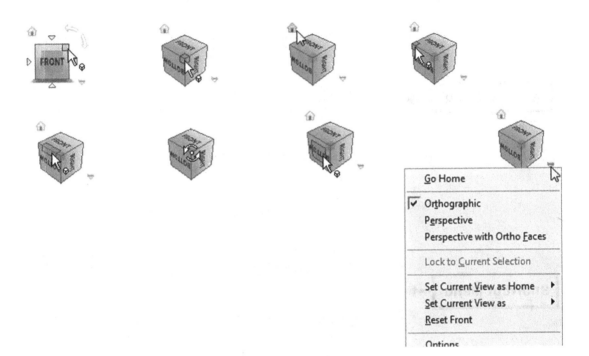

Shortcut Menus and Marking Menus

When you right-click, a shortcut menu along with a marking menu appears. A shortcut menu contains a list of some important options. The marking menu contains important tools. It allows you to access the tools quickly. You can customize the marking menu by adding and removing tools.

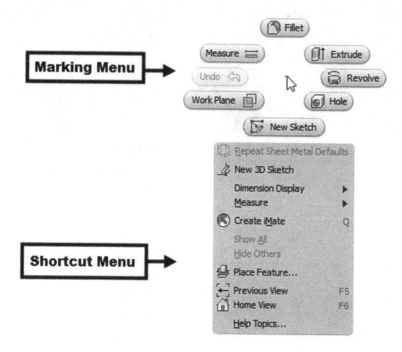

Dialogs

When you activate any tool in Autodesk Inventor, the dialog related to it appears. The dialog consists of various options that help you to complete the operation. The following figure shows the components of a dialog:

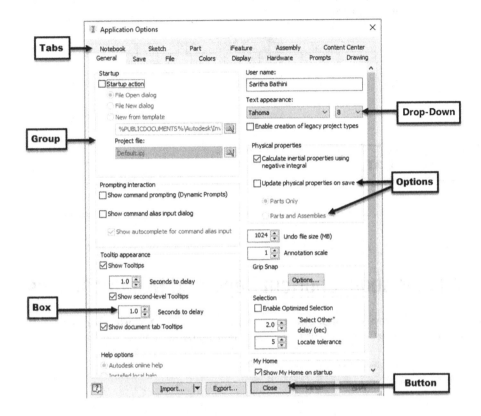

Using the Mini-Toolbar

The mini-toolbar appears in the dialog boxes of the Extrude, Revolve, Fillet, Shell, Face Draft, Chamfer, and Joint commands. However, in Autodesk Inventor 2018, the mini-toolbar does not appear by default. You need to check the **Mini-Toolbar** option available in the **User Interface** drop-down of the **Windows** panel of the View ribbon tab to display the mini-toolbar.

Customizing the Ribbon, Shortcut Keys, and Marking Menus

To customize the ribbon, shortcut keys, or marking menus, click **Tools ➤ Options ➤ Customize** on the ribbon. In the **Customize** dialog, use the tabs to customize the ribbon, marking menu, or shortcut keys.

For example, to add a command to the ribbon, select the command from the list on the left side of the dialog and click the **Add** button. If you want to remove a command from the ribbon, then select it from the right-side list and click the **Remove** button. Click **OK** to make the changes take effect.

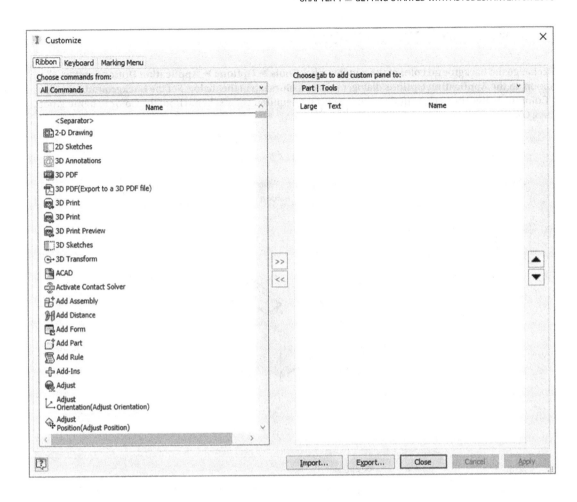

To add or remove panels from the ribbon, click the **Show Panels** icon located on the right side of the ribbon and check/uncheck the options on the fly-out menu.

Exploring the Color Settings

To change the background color of the window, click **Tools ➤ Options ➤ Application Options** on the ribbon. In the **Application Options** dialog, click the **Colors** tab in the dialog. Set the **Background** value to **1 Color** to change the background to plain. Select the required color scheme from the **Color Scheme** group. Click **OK**.

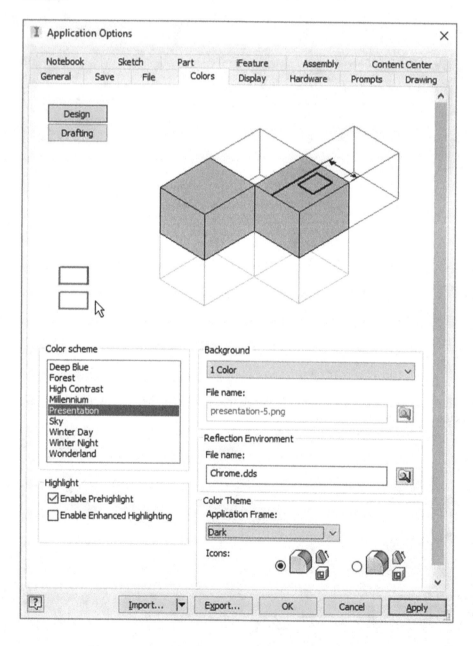

CHAPTER 2

■ ■ ■

Part Modeling Basics

This chapter takes you through the creation of your first Inventor model. You will create simple parts. In this chapter, you will do the following:

- Create sketches
- Create a base feature
- Add another feature to it
- Create revolved features
- Create cylindrical features
- Create box features
- Apply draft

Tutorial 1

This tutorial takes you through the creation of your first Inventor model. You will create the disc of an Oldham coupling.

© T. Kishore 2017
T. Kishore, *Learn Autodesk Inventor 2018 Basics*, https://doi.org/10.1007/978-1-4842-3225-5_2

Creating a New Project

Follow these steps:

1. Start **Autodesk Inventor 2018** by double-clicking the **Autodesk Inventor 2018** icon on your desktop.

2. To create a new project, click **Get Started ➤ Launch ➤ Projects** on the ribbon.

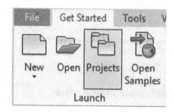

3. Click the **New** button in the **Projects** dialog.

4. In the **Inventor project wizard** dialog, select **New Single User Project** and click the **Next** button.

5. Enter **Oldham Coupling** in the **Name** field.

6. Enter **C:\Users\Username\Documents\Inventor\Oldham Coupling** in the **Project (Workspace) Folder** box and click **Next**.

7. Click **Finish**.

8. Click **OK** in the **Inventor Project Editor** dialog.

9. Click **Done**.

Starting a New Part File

Follow these steps:

1. To start a new part file, click **Get Started ➤ Launch ➤ New** on the ribbon.

2. In the **Create New File** dialog, click the **Templates** folder located in the top-right corner.

3. Click the **Standard.ipt** icon located in the **Part – Create 2D and 3D Objects** section.

4. Click the **Create** button in the **Create New File** dialog.

A new model window appears.

Starting a Sketch

Follow these steps:

1. To start a new sketch, click **3D Model ➤ Sketch ➤ Start 2D Sketch** on the ribbon.

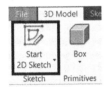

2. Click the **XY plane**. The sketch starts.

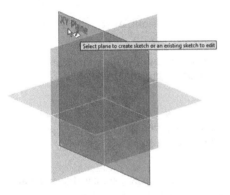

> The first feature is an extruded feature from a sketched circular profile. You will begin by sketching the circle.

3. On the ribbon, click **Sketch ➤ Create ➤ Circle ➤ Circle Center Point**.

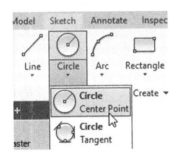

4. Move the cursor to the sketch origin located at the center of the graphics window and then click it.

5. Drag the cursor up to a random location and then click to create a circle.

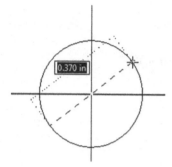

6. Press **Esc** to deactivate the tool.

Adding Dimensions

In this section, you will specify the size of the sketched circle by adding dimensions. As you add dimensions, the sketch can attain any one of the following states:

- *Fully constrained sketch*: In a fully constrained sketch, the positions of all the entities are fully described by dimensions, constraints, or both. In a fully constrained sketch, all the entities are dark blue color.

- *Under-constrained sketch*: Additional dimensions, constraints, or both are needed to completely specify the geometry. In this state, you can drag under-constrained sketch entities to modify the sketch. An under-constrained sketch entity is displayed in black.

If you add any more dimensions to a fully constrained sketch, a message box will appear showing that dimension over constrains the sketch. In addition, it prompts you to convert the dimension into a driven dimension. Click **Accept** to convert the unwanted dimension into a driven dimension.

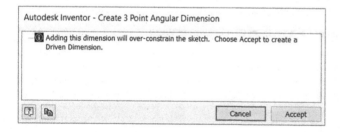

1. On the ribbon, click **Sketch ➤ Constrain ➤ Dimension**.

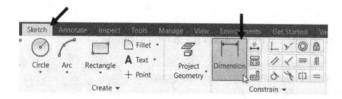

2. Select the circle and click; the **Edit Dimension** box appears.

3. Enter **4** in the **Edit Dimension** box and click the green check.

4. Press **Esc** to deactivate the **Dimension** tool.

   You can also create dimensions while creating the sketch objects. To do this, enter the dimension values in the boxes displayed while sketching.

5. To display the entire circle at full size and to center it in the graphics area, use one of the following methods:

   • Click **Zoom All** on the **Navigation Bar**.

   • Click **View ➤ Navigate ➤ Zoom All** on the ribbon.

6. Click **Finish Sketch** on the **Exit** panel.

7. Again, click **Zoom All** on the **Navigation Bar**.

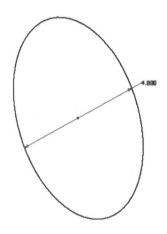

Creating the Base Feature

The first feature in any part is called a *base feature*. You now create this feature by extruding the sketched circle.

1. On the ribbon, click **3D Model ➤ Create ➤ Extrude**.

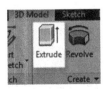

2. Type **0.4** in the **Distance** box available in the **Extrude** dialog.

3. Click the **Direction 1**  icon in the **Extrude** dialog.

4. Click **OK** in the **Extrude** dialog to create the extrusion.

Notice the new feature, **Extrusion 1**, in the **Browser window**.

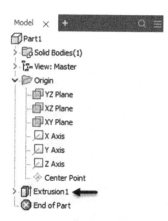

To magnify a model in the graphics area, you can use the zoom tools available in the **Zoom** drop-down in the **Navigate** panel of the **View** tab.

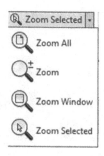

Click **Zoom All** to display the part full-size in the current window.

Click **Zoom Window** and then drag the pointer to create a rectangle; the area in the rectangle zooms to fill the window.

Click **Zoom** and then drag the pointer. Dragging up zooms out; dragging down zooms in.

Click a vertex, an edge, or a feature and then click **Zoom Selected**; the selected item zooms to fill the window.

To display the part in different rendering modes, select the options in the **Visual Style** drop-down on the **Appearance** panel of the **View** tab. The default display mode for parts and assemblies is **Shaded with Edges**. You may change the display mode whenever you want.

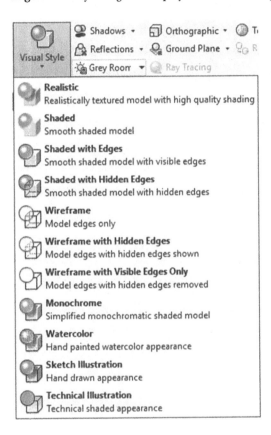

Realistic

Shaded

Shaded with Edges

Shaded with Hidden Edges

Wireframe

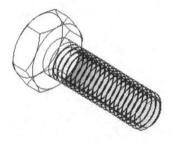

Wireframe with Hidden Edges

Wireframe with Visible Edges Only

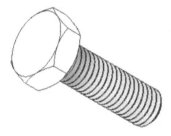

Monochrome

Watercolor

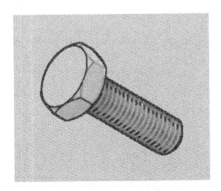

Sketch Illustration

Technical Illustration

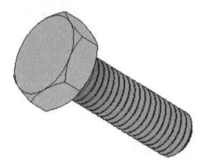

Adding an Extruded Feature

To create additional features on the part, you need to draw sketches on the model faces or planes and then extrude them.

1. On the ribbon, click **View ➤ Appearance ➤ Visual Style ➤ Wireframe**.

2. On the ribbon, click **3D Model ➤ Sketch ➤ Start 2D Sketch**.

3. Click the front face of the part.

4. Click **Line** on the **Create** panel.

5. Click the circular edge to specify the first point of the line.

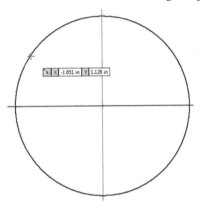

6. Move the cursor toward the right.

7. Click the other side of the circular edge; a line is drawn.

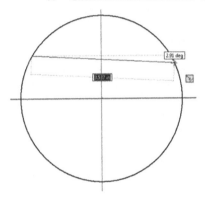

8. Draw another line below the previous line.

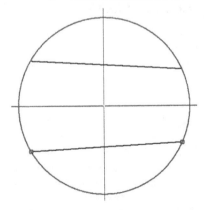

9. On the ribbon, click **Sketch ➤ Constrain ➤ Horizontal Constraint** .

10. Select the two lines to make them horizontal.

11. On the ribbon, click **Sketch ➤ Constrain ➤ Equal** $=$.

12. Select the two horizontal lines to make them equal.

13. Click **Dimension** on the **Constrain** panel of the **Sketch** ribbon tab.

14. Select the two horizontal lines.

15. Move the cursor toward the right and click to locate the dimension; the **Edit Dimension** box appears.

16. Enter **0.472** in the **Edit Dimension** box and click the green check.

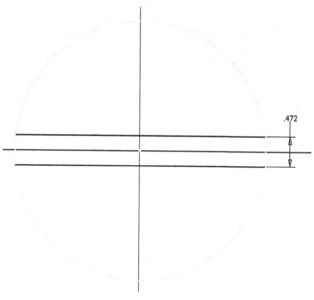

17. Click **Finish Sketch** on the **Exit** panel.

18. Click the sketch and then click **Create Extrude** on the **mini-toolbar**; the **Extrude** dialog appears.

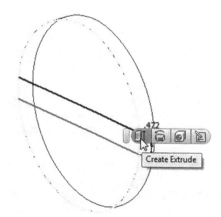

19. Click in the region bounded by the two horizontal lines.

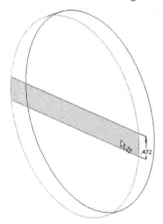

20. Enter **0.4** in the **Distance1** box in the **Extrude** dialog.

21. In the **Extrude** dialog, click the **Direction 1** icon and then click **OK** to create the extrusion.

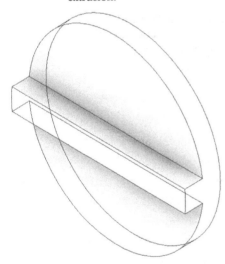

By default, the ambient shadows are displayed on the model. However, you can turn off the ambient shadows by clicking the **View** tab, going to the **Appearance** panel, selecting the **Shadows** drop-down, and then deselecting the **Ambient Shadows** option. The **Shadows** drop-down has two more options, which you use based on your requirements.

You can reuse the sketch of an already existing feature. To do this, expand the feature in the Browser window, right-click the sketch, and select **Share Sketch** from the shortcut menu. You will notice that the sketch is visible in the graphics window. You can also unshare the sketch by right-clicking it and selecting **Unshare**.

Adding Another Extruded Feature

Follow these steps:

1. Click **Start 2D Sketch** on the **Sketch** panel of the **3D Model** ribbon tab.

2. Use the **Free Orbit** button on the **Navigation Bar** to rotate the model so that the back face of the part is visible.

3. Right-click and select **OK**.

4. Click the back face of the part.

5. Click **Line** on the **Create** panel.

6. Draw two lines, as shown next (refer to the "**Adding an Extruded Feature**" section for how to draw lines). Make sure that the endpoints of the lines are coincident with the circular edge. Follow the next two steps, if they are not coincident.

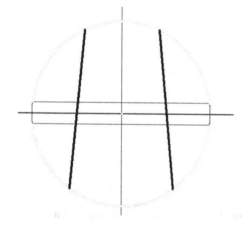

7. On the ribbon, click **Sketch ➤ Constrain ➤ Coincident Constraint** ⌊▬ . Next, select the endpoint of the line and the circular edge.

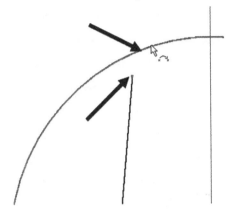

8. Likewise, make the other endpoints of the lines coincident with the circular edge.

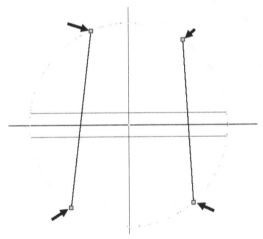

Skip the previous two steps if the endpoints of the lines are coincident with the circular edge.

📝 You can specify a point using various point snap options. To do this, activate a sketching tool, right-click, and select **Point Snaps**; a list of point snaps appears. Now, you can select only the specified point snap.

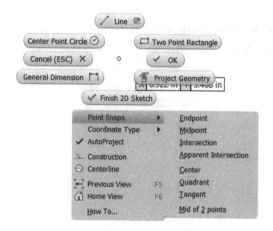

9. On the ribbon, click **Sketch ➤ Constrain ➤ Vertical Constraint** ⫴ .

10. Select the two lines to make them vertical.

11. On the ribbon, click **Sketch ➤ Constrain ➤ Equal** = .

12. Select the two vertical lines to make them equal.

13. Create a dimension of 0.472 between the vertical lines.

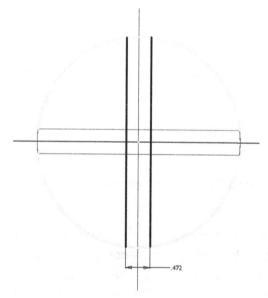

14. Click **Finish Sketch**.

15. On the ribbon, click **3D Model ➤ Create ➤ Extrude**.

16. Click inside the region enclosed by two lines, if they are not already selected.

17. Type **0.4** in the **Distance1** box in the **Extrude** dialog and click **OK**.

To move the part view, click **Pan** 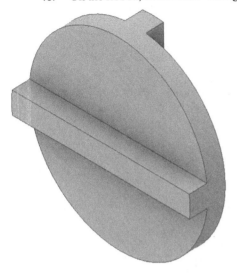 on the **Navigation Bar** and then drag the part to move it in the graphics area.

18. On the ribbon, click **View ➤ Appearance ➤ Visual Style ➤ Shaded with Edges**.

19. On the ribbon, click **View ➤ Navigate ➤ Home View** ⌂.

Saving the Part

Follow these steps:

1. Click **Save** 🖫 on the Quick Access Toolbar.

2. In the **Save As** dialog, type **Disc** in the **File name** box.

3. Click **Save** to save the file.

4. Click **File ➤ Close**.

Note that *.ipt is the file extension for all the files you create in the Part environment of Autodesk Inventor.

Tutorial 2

In this tutorial, you will create a flange by performing the following:

- Creating a revolved feature
- Creating a cut features
- Adding fillets

Starting a New Part File

Follow these steps:

1. To start a new part file, click the **Part** icon on the Home screen.

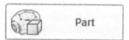

Sketching a Revolve Profile

You create the base feature of the flange by revolving a profile around a centerline.

1. Click **3D Model ➤ Sketch ➤ Start 2D Sketch** on the ribbon.

2. Select the YZ plane.

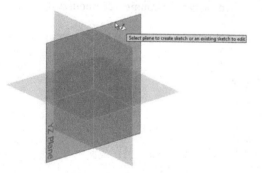

3. Click **Line** 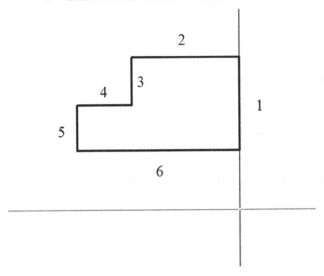 on the **Create** panel.

4. Create a sketch similar to the one shown here:

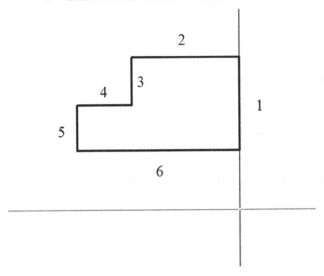

5. On the ribbon, click **Sketch ➤ Format ➤ Centerline** .

6. Click **Line** on the **Create** panel.

7. Create a centerline, as shown here:

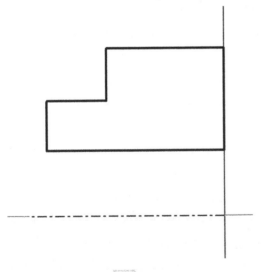

8. Click **Fix** on the **Constrain** panel.

9. Select Line 1.

10. Click **Dimension** on the **Constrain** panel.

33

11. Select the centerline and Line 2; a dimension appears.

12. Move the pointer horizontally toward the right and click to place the dimension.

13. Place the dimension and enter **4** in the **Edit Dimension** box.

14. Click the green check ✔ in the **Edit Dimension** dialog.

15. Select the centerline and Line 4; a dimension appears.

16. Move the pointer horizontally toward the left and click to place the dimension.

17. Enter **2.4** in the **Edit Dimension** box.

18. Click the green check ✔ in the **Edit Dimension** dialog.

19. Select the centerline and Line 6; a dimension appears.

20. Move the pointer horizontally toward the left and click to place the dimension.

21. Enter **1.2** in the **Edit Dimension** box.

22. Click the green check ✔ in the **Edit Dimension** dialog.

23. Create a dimension between the Line 1 and Line 3.

24. Set the dimension value to 0.8 inches.

25. Create a dimension between Line 1 and Line 5.

26. Set the dimension value to 2 inches.

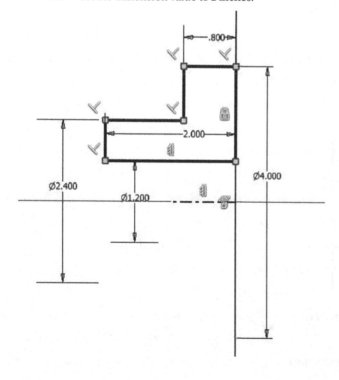

You can display all the constraints by right-clicking and selecting the **Show All Constraints** option. You can hide all the constraints by right-clicking and selecting the **Hide All Constraints** option.

27.   Right-click and select **Finish 2D Sketch**.

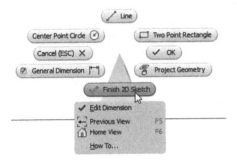

Creating the Revolved Feature

Follow these steps:

1.   On the ribbon, click **3D Model ➤ Create ➤ Revolve** or right-click and select **Revolve** from the marking menu.

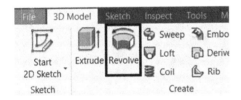

2.   Set **Extents** to **Full** in the **Revolve** dialog.

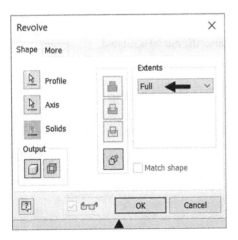

3. Click **OK** to create the revolved feature.

Creating the Cut Feature

Follow these steps:

1. On the Navigation Bar, click the **Orbit** icon.

2. Press and hold the left mouse button and drag the mouse; the model is rotated.

3. Rotate the model so that its back face is visible.

4. Right-click and select **OK**.

5. On the **3D Model** tab of the ribbon, click the **Show Panels** icon located in the right corner and then select the **Primitives** option from the drop-down.

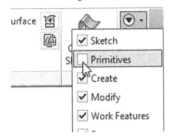

The **Primitives** panel is added to the ribbon.

6. On the ribbon, click **3D Model**, go to the **Primitives** panel, click the **primitive drop-down**, and select Box.

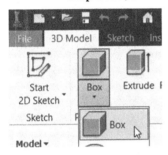

7. Click the back face of the part; the sketch starts.

8. Select the origin to define the center point.

9. Move the cursor diagonally toward the right.

10. Enter **4.1** in the horizontal dimension box.

11. Press the Tab key and enter **0.472** in the vertical dimension box.

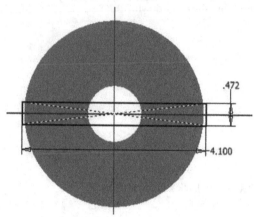

12. Press the Enter key; the **Extrude** dialog appears.

13. Expand the **Extrude** dialog by clicking the down arrow ▼ button.

14. Click the **Cut** 🖮 button in the **Extrude** dialog.

15. Enter **0.4** in the **Distance** box.

16. Click **OK** to create the cut feature.

Creating Another Cut Feature

Follow these steps:

1. Click the Home icon located at the top-left corner of the ViewCube.

2. Create a sketch on the front face of the base feature.

   - On the ribbon, click **3D Model ➤ Sketch ➤ Start 2D Sketch**.

   - Select the front face of the model.

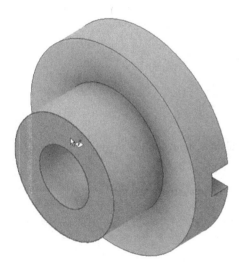

- Draw three lines and the circle, as shown here:

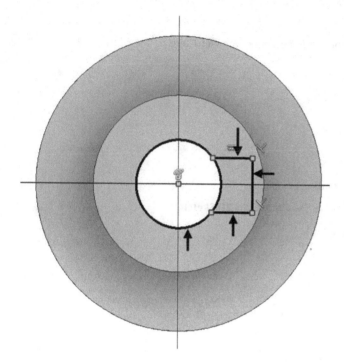

- Apply the **Horizontal** constraint to the horizontal lines, if not applied already.
- Apply the **Equal** constraint between the horizontal lines.
- Ensure that the endpoints of the horizontal line coincide with the circle.
- Apply a dimension of 0.236 to the vertical line.
- Apply a dimension of 0.118 to the horizontal line.

- Apply a dimension with a 1.2 diameter to the circle.

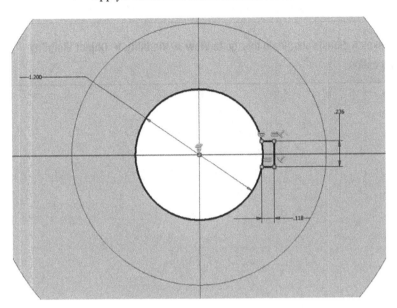

- On the ribbon, click **Sketch ➤ Modify ➤ Trim** .
- Click the circle to trim it.

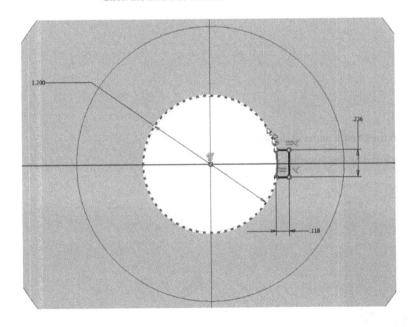

3. Finish the sketch.

 You can hide or display the sketch dimensions. To do this, go to **View ➤ Visibility ➤ Object Visibility** and check the **Sketch Dimensions** option.

4. Click **Extrude** 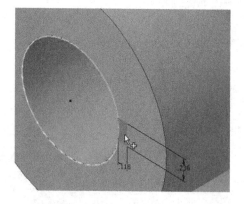 on the **Create** panel of the **3D model**.
5. Click in the region enclosed by the three lines and the arc.

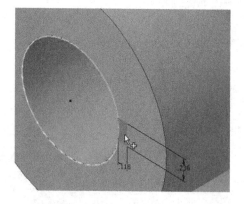

6. Select **All** from the **Extents** drop-down.

7. Click the **Cut** 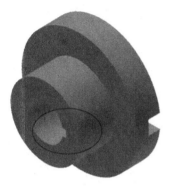 button in the **Extrude** dialog.

8. Click **OK** to create the cut feature.

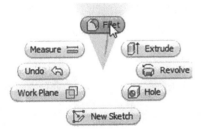

Adding a Fillet

Follow these steps:

1. On the ribbon, click **3D Model ➤ Modify ➤ Fillet** or right-click and select **Fillet** from the marking menu.

2. Click the inner circular edge and set **Radius** as 0.2.

3. Click **OK** to add the fillet.

Saving the Part

Follow these steps:

1. Click **Save** on the **Quick Access Toolbar**.
2. In the **Save As** dialog, type-in **Flange** in the **File name** box.
3. Click **Save** to save the file.
4. Click **File ➤ Close**.

Tutorial 3

In this tutorial, you will create a shaft by performing the following:

- Creating a cylindrical feature
- Creating a cut feature

Starting a New Part File

Follow these steps:

1. On the ribbon, click **Get Started ➤ Launch ➤ New** 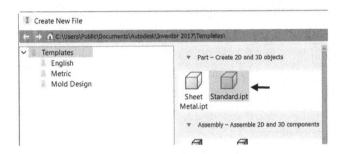.

2. In the **Create New File** dialog, select **Standard.ipt**.

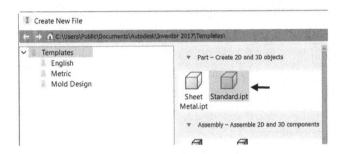

3. Click **Create**.

Creating the Cylindrical Feature

Follow these steps:

1. On the ribbon, click **Primitives**, click the **primitive** drop-down, and select **Cylinder**.

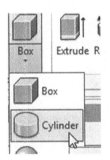

2. Click the XY plane to select it; the sketch starts.

3. Click at the origin and move the cursor outward.

4. Enter **1.2** in the box attached to the circle.

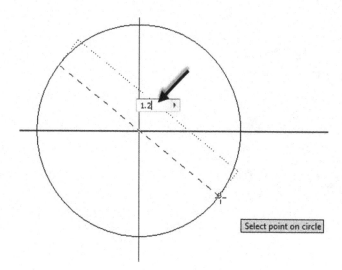

Select point on circle

5. Press the Enter key; the **Extrude** dialog appears.

6. Enter **4** in the **Distance** box.

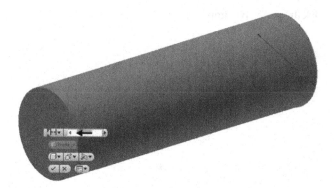

7. Click **OK** to create the cylinder.

Creating Cut Feature

Follow these steps:

1. Create a sketch on the front face of the base feature.

   • On the ribbon, click **3D Model ➤ Sketch ➤ Start 2D Sketch**.

   • Select the front face of the cylinder.

   • On the ribbon, click **Sketch ➤ Create ➤ Line**.

- Draw three lines, as shown here:

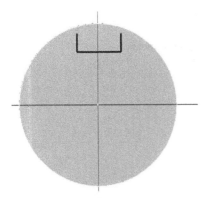

- Apply the **Coincident** constraint between the endpoints of the vertical lines and the circular edge.

- Add dimensions to the sketch.

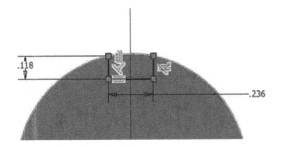

2. Finish the sketch.

3. Click **Extrude** in the **Create** panel.

4. Click in the region enclosed by the sketch.

5. Click the **Cut** button in the **Extrude** dialog.

6. Set **Distance** to **2.165**.

7. Click **OK** to create the cut feature.

Saving the Part

Follow these steps:

1. Click **Save**  on the **Quick Access Toolbar**; the **Save As** dialog appears.
2. Type **Shaft** in the **File name** box.
3. Click **Save** to save the file.
4. Click **File ➤ Close**.

Tutorial 4

In this tutorial, you will create a key by performing the following:

- Creating an extruded feature
- Applying draft

Starting an Extruded Feature

Follow these steps:

1. Start a new part file using the **Standard.ipt** template.
2. On the ribbon, select **Primitives**, click the **primitive** drop-down, and select **Box**.
3. Select the XY plane.
4. Create the sketch, as shown here:

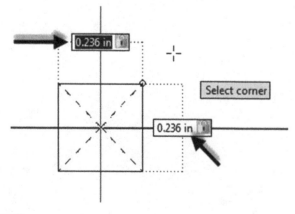

5. Press Enter.

6. Enter **2** in the **Distance** box.

7. Click **OK** to create the extrusion.

Applying Draft

Follow these steps:

1. On the ribbon, click **3D Model ➤ Modify ➤ Draft**.

2. Select the **Fixed Plane** option.

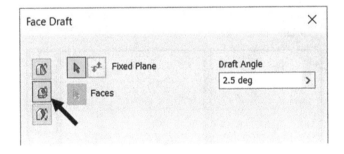

3. Select the front face as the fixed face.

4. Select the top face as the face to be draft.

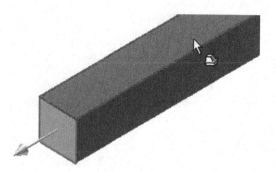

5. Set **Draft Angle** to **1**.

6. Click the **Flip Pull Direction** button in the **Face Draft** dialog.

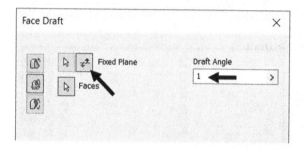

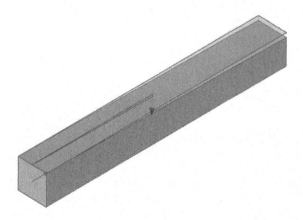

7. Click **OK** to create the draft.

Saving the Part

Follow these steps:

1. Click **Save** on the **Quick Access Toolbar**; the **Save As** dialog appears.

2. Type **Key** in the **File name** box.

3. Click **Save** to save the file.

4. Click **File ➤ Close**.

CHAPTER 3

■ ■ ■

Assembly Basics

In this chapter, you will do the following:

- Add components to an assembly
- Apply constraints between components
- Check the degrees of freedom
- Check the interference
- Create an exploded view of the assembly

Tutorial 1

This tutorial takes you through the creation of your first assembly. You will create the Oldham coupling assembly shown here:

PARTS LIST		
ITEM	PART NUMBER	QTY
1	Disc	1
2	Flange	2
3	Shaft	2
4	Key	2

© T. Kishore 2017
T. Kishore, *Learn Autodesk Inventor 2018 Basics*, https://doi.org/10.1007/978-1-4842-3225-5_3

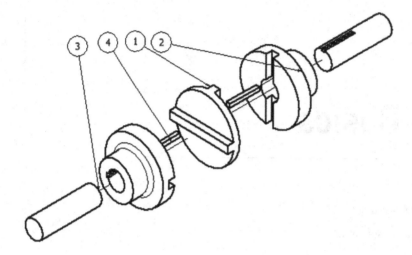

There are two ways of creating any assembly model.

- Top-down approach
- Bottom-up approach

Top-Down Approach

In a top-down approach, the assembly file is created first, and components are created in that file.

Bottom-Up Approach

In a bottom-up approach, the components are created first and then added to the assembly file. In this tutorial, you will create the assembly using this approach.

Starting a New Assembly File

To start a new assembly file, click the **Assembly** icon on the Home screen.

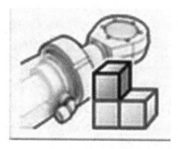

Inserting the Base Component

Follow these steps:

1. To insert the base component, click **Assemble ➤ Component ➤ Place** on the ribbon.

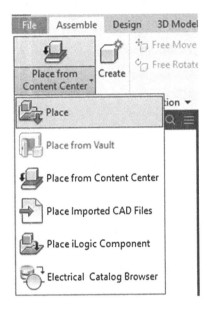

2. Browse to the project folder and double-click **Flange.ipt**.

3. Right-click and select **Place Grounded at Origin**; the component is placed at the origin.

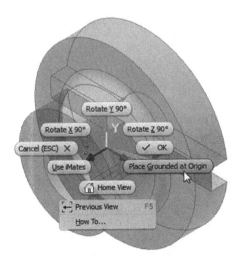

4. Right-click and select **OK**.

Adding the Second Component

Follow these steps:

1. To insert the second component, right-click and select **Place Component**; the **Place Component** dialog appears.

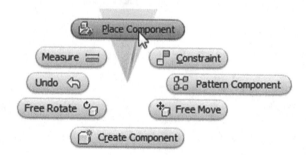

2. Browse to the project folder and double-click **Shaft.ipt**.

3. Click in the window to place the component.

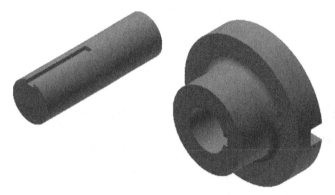

4. Right-click and select **OK**.

Applying Constraints

After adding the components to the assembly environment, you need to apply constraints between them. By applying constraints, you establish relationships between components.

1. To apply constraints, click **Assemble** ➤ **Relationships** ➤ **Constrain** on the ribbon.

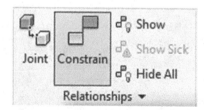

The **Place Component** dialog appears on the screen.

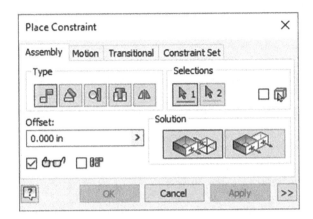

Different assembly constraints that can be applied using this dialog are given next. Using the **Mate** constraint, you can make two planar faces coplanar to each other.

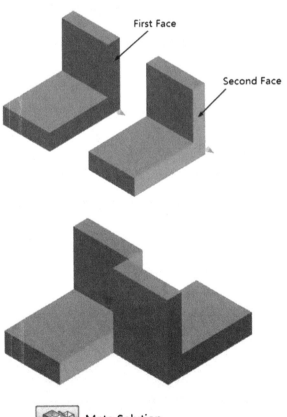

Note that if you set the **solution** to **Flush**, the faces will point in the same direction.

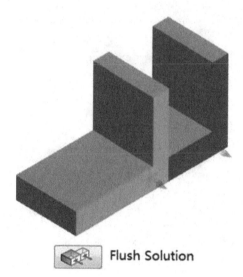

Flush Solution

You can also align the centerlines of the cylindrical faces.

Angle applies the angle constraint between two components.

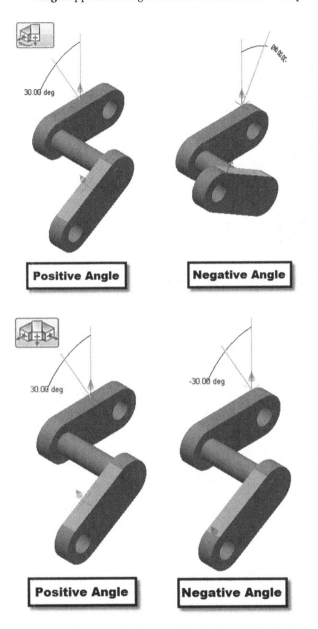

The **Tangent** constraint is used to apply a tangent relation between two faces.

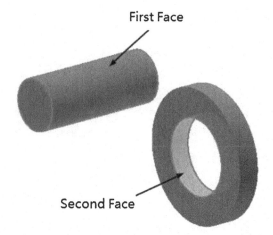

 Outside Solution

Inside Solution

The **Insert** constraint is used to make two cylindrical faces coaxial. In addition, the planar faces of the cylindrical components will be on the same plane.

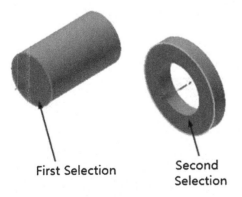

First Selection

Second
Selection

 Opposed Solution

 Aligned Solution

The **Symmetry** constraint is used to position the two components symmetrically about a plane.

First Face

Second Face

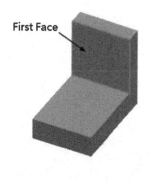
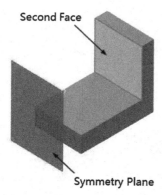

Symmetry Plane

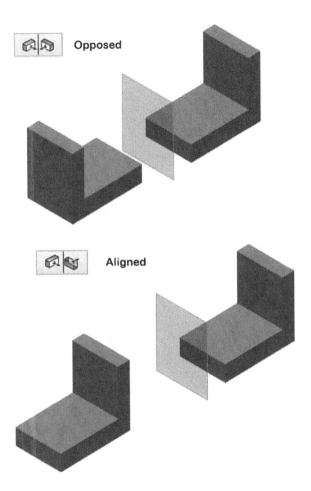

1. In the **Place Constraints** dialog, under the **Type** group, click the **Mate** 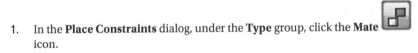 icon.

2. Click the cylindrical face of the shaft.

3. Click the inner cylindrical face of the flange.

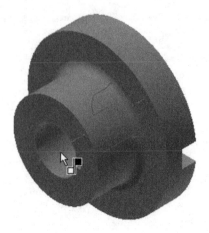

4. Click the **Apply** button.

5. Ensure that the **Mate**  icon is selected in the **Type** group.

6. Click the front face of the shaft.

7. Rotate the model.

8. Click the slot face of the flange, as shown here:

9. Click the **Flush** button in the **Place Constraint** dialog.

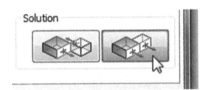

10. Click **Apply**. The front face of the Shaft and the slot face of the flange are aligned.

11. Ensure that the **Mate** button is selected in the **Type** group.

12. Expand **Flange: 1** in the Browser window.

13. Select the XZ plane of the flange.

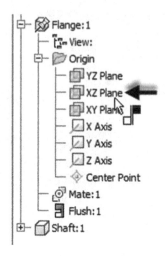

14. Expand **Shaft: 1** and select the YZ plane of the shaft.

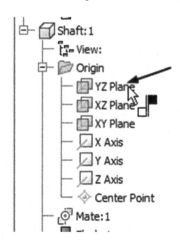

15. Click the **Flush** button in the **Place Constraint** dialog.

16. Click **OK** to assemble the components.

Adding the Third Component

Follow these steps:

1. To insert the third component, click **Assemble ➤ Component ➤ Place** on the ribbon.

2. Go to the project folder and double-click **Key.ipt**.

3. Click in the graphics window to place the key.

4. Right-click and select **OK**.

5. Right-click **Flange: 1** in the Browser window.

6.   Click **Visibility** on the shortcut menu; the flange is hidden.

7.   Click **Constrain** 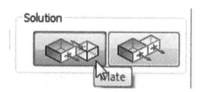 on the **Relationships** panel.

8.   Click **Mate** in the **Place Constraint** dialog.

9.   Select **Mate** from the **Solution** group.

10. Right-click the side face of the key and **Select Other** on the shortcut menu.

11. Select the bottom face of the key from the fly-out menu.

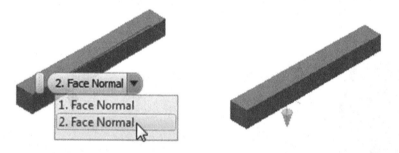

12. Select the flat face of the slot.

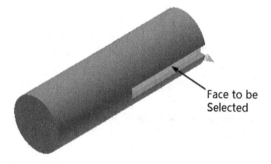

Face to be
Selected

13. Click the **Apply** button. The bottom face of the key is aligned with the flat face of the slot.

14. Click the **Mate** icon in the **Place Constraint** dialog.

15. Select **Flush** from the **Solution** group.

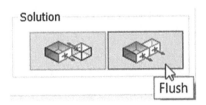

16. Select the front face of the key and back face of the shaft, as shown here:

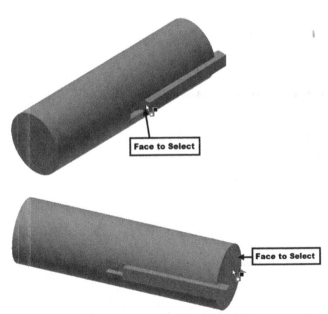

17. Click **Apply** in the dialog; the mate is applied.

18. Close the dialog.

   Now, you need to check whether the parts are fully constrained or not.

19. Click **View ➤ Visibility ➤ Degrees of Freedom** on the ribbon.

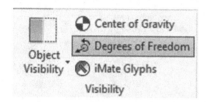

You will notice that an arrow appears pointing in the upward (or downward) direction. This means that the key is not constrained in the Z-direction.

   You must apply one more constraint to constrain the key.

20. Click **Constrain** on the **Relationships** panel of the **Assemble** ribbon tab.

21. Click the **Mate** icon in the dialog.

22. Select **Flush** from the **Solution** group.

23. Expand the **Origin** node of the **assembly** in the **Browser window** and select **XZ Plane**.

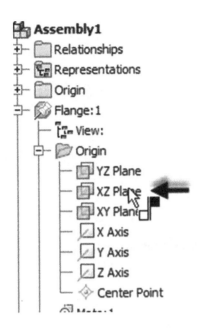

24.   Expand the **Key: 1** node in the **Browser window** and select **YZ Plane**.

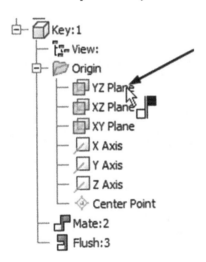

25.   Click **OK**. The mate is applied between the two planes.

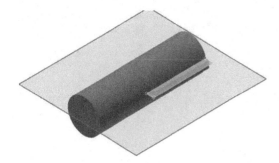

Now, you need to turn on the display of the flange.

26.   Right-click the **flange** in the **Browser window** and select **Visibility**; the **flange** appears.

Checking the Interference

Follow these steps:

1.   Click **Inspect ➤ Interference ➤ Analyze Interference** on the ribbon. The **Interference Analysis** dialog appears.

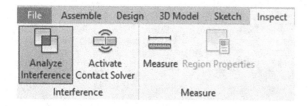

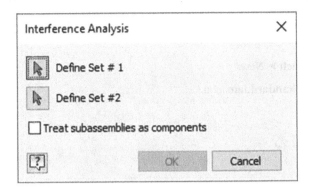

2. Select the flange and shaft as **Set #1**.

3. Click the **Define Set #2** button.

4. Select the key as **Set #2**.

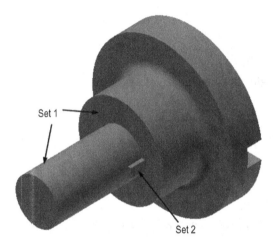

5. Click **OK**; the message box appears showing that there are no interferences.

6. Click **OK**.

Saving the Assembly

Follow these steps:

1. Click **Save** on the **Quick Access Toolbar**; the **Save As** dialog appears.

2. Type **Flange_subassembly** in the **File name** box.

3. Go to the project folder.

4. Click **Save** to save the file.

5. Click **File ➤ Close** .

Starting the Main Assembly

Follow these steps:

1. On the ribbon, click **Get Started ➤ Launch ➤ New**.

2. In the **Create New File** dialog, click the **Standard.iam** icon.

Standard.iam

3. Click **Create** to start a new assembly.

Adding a Disc to the Assembly

Follow these steps:

1. Click **Assemble ➤ Component ➤ Place** on the ribbon.

2. Go to the project folder and double-click **Disc.ipt**.

3. Right-click and select **Place Grounded at Origin**; the component is placed at the origin.

4. Right-click and select **OK**.

Placing the Subassembly

Follow these steps:

1. To insert the subassembly, click the **Place** button on the **Component** panel of the ribbon.

2. Go to the project folder and double-click **Flange_subassembly.iam**.

3. Click in the window to place the flange subassembly.

4. Right-click and select **OK**.

Adding Constraints

Follow these steps:

1. Click **Constrain** on the **Relationships** panel of the **Assemble** ribbon.

2. Click the **Insert** button in the **Place Constraint** dialog.

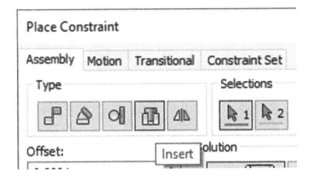

3. Select **Opposed** from the **Solution** group.

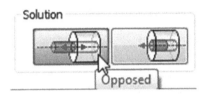

4. Click the circular edge of the flange.

5. Click the circular edge of the disc.

6. Click **OK** in the dialog.

    Next, you have to move the subassembly away from the disc to apply other constraints.

7. Click **Free Move** on the **Position** panel.

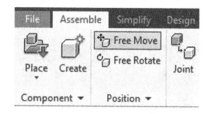

8. Select the flange subassembly and move it.

9. Click the **Constrain** button on the **Relationships** panel.

10. Click **Mate** in the **Place Constraints** dialog.

11. Select **Mate** from the **Solution** group.

12. Click **View ➤ Navigate ➤ Orbit** on the ribbon.

13. Press and hold the left mouse button and drag the cursor toward the left.

14. Release the mouse button, right-click, and select **OK**.

15. Click the face on the flange, as shown here:

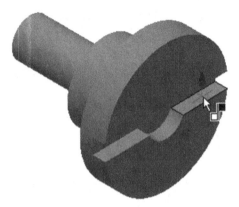

16. Click the face on the disc, as shown here:

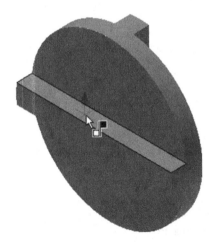

17. Click **OK** in the dialog.

Placing the Second Instance of the Subassembly

Follow these steps:

1. Insert another instance of the flange subassembly.

2. Apply the **Insert** and **Mate** constraints.

Saving the Assembly

Follow these steps:

1. Click **Save** on the **Quick Access Toolbar**; the **Save As** dialog appears.

2. Type **Oldham_coupling** in the **File name** box.

3. Click **Save** to save the file.

4. Click **File ➤ Close**.

Tutorial 2

In this tutorial, you will create the exploded view of the assembly, as shown here:

Starting a New Presentation File

Follow these steps:

1. On the Home screen, click the **Presentation** icon or click **Get Started ➤ Launch ➤ New** and then select the **Standard.ipn** template from the **Create New File** dialog.

The **Insert** dialog appears.

2. In the **Insert** dialog, go to the project folder and double-click the **Oldham Coupling.iam** file.

The Presentation environment appears, as shown here:

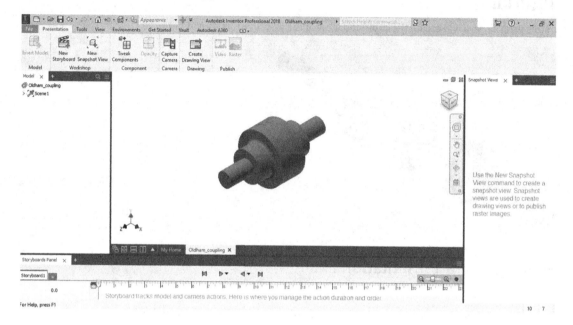

Creating a Storyboard Animation

Follow these steps:

1. In the **Model** tree, double-click **Scene1** and type **Explosion**.

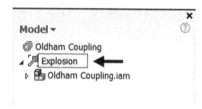

Before creating an exploded view, you need to take a look at the storyboard displayed at the bottom of the window. The storyboard has the Scratch Zone area located at the left side of the timeline. Also, notice that the play marker is displayed at 0 seconds in the timeline.

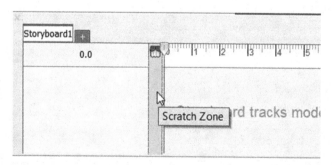

2. Click in the Scratch Zone area and notice that the play marker is displayed inside it.

   Now, the changes made to the assembly in the Scratch Zone area will be the starting point of the exploded animation. You can change the orientation of the assembly, hide a component, or change the opacity of the component. Use the **Capture Camera** tool on the **Camera** panel to set the camera position for the animation.

3. On the ribbon, click the **View** tab, go to the **Windows** panel, click the **User Interface** drop-down, and then check the **Mini-Toolbar** option. The mini-toolbar appears whenever you activate a tool.

4. Click the **Tweak Components** button on the **Component** panel of the **Presentation** ribbon tab. The mini-toolbar appears with different options, as shown here:

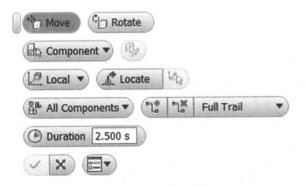

Notice that the default duration for a tweak is 2.500 s. You can type a new value in the **Duration** box available on the mini-toolbar.

5. Select **Component** from the Selection Filter drop-down of the mini-toolbar.

6. Select **All Components** from the Tracelines drop-down. This will create tracelines of all exploded components.

7. Select the flange subassembly from the graphics window. The manipulator appears on the assembly.

Now, you must specify the direction along which the subassembly will be exploded.

8. On the mini-toolbar, select **Local ➤ World**.

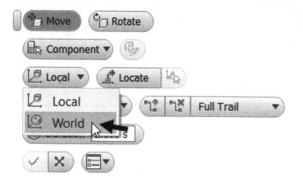

9. Click the z-axis of the manipulator.

10.   Type **4** in the Z box attached to the manipulator.

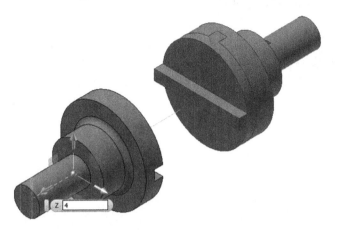

11.   Click **OK** on the mini-toolbar.

12.   Right-click in the graphics window and select **Tweak Components** from the marking menu.

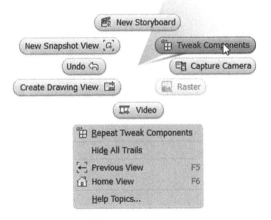

13.   Select **Component** from the Selection Filter drop-down on the mini-toolbar.

14.   Select the other flange subassembly.

15.   Click the z-axis of the manipulator.

16.   Type **4** in the Z box attached to the manipulator and click **OK**

17.   Click the **Tweak Components** button on the **Component** panel of the **Presentation** ribbon tab.

18.   On the mini-toolbar, select **Part** from the drop-down, as shown here:

19.   Select the front cylinder.

20.   On the mini-toolbar, select **Local ➤ World**.

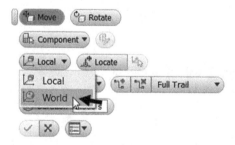

21.　Click the z-axis of the manipulator.

22.　Type **4** in the box attached to the manipulator and then press Enter.

23.　Click **OK** on the mini-toolbar.

24.　Activate the **Tweak Components** command.

25.　Zoom into the flange and click the key, as shown here:

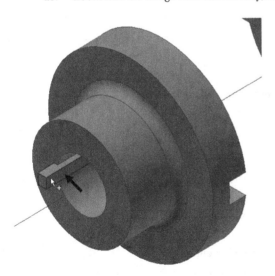

26.　Click the z-axis of the manipulator.

27.　Type **3.15** in the box attached to the manipulator and press Enter.

28. Likewise, explode the parts of the flange subassembly in the opposite direction. The explosion distances are same.

Animating the Explosion

Follow these steps:

1. To play an animation of the explosion, click the **Play Current Storyboard** button on the storyboard.

2. Click the **Reverse Play Current Storyboard** button on the storyboard.

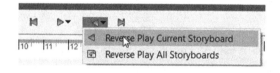

You can publish the animation video using the **Video** tool available on the **Publish** panel.

3. Make sure that the play marker is at 0 seconds on the timeline.

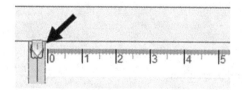

4. On the ribbon, click **Presentation ➤ Publish ➤ Video** 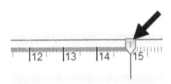. Make sure that the play marker is at 0 seconds on the timeline.

5. In the **Publish to Video** dialog, select the **Current Storyboard** option from the **Publish Scope** section. Make sure the play marker is at 0 seconds on the timeline.

    You can also select **Current Storyboard Range** and specify the start and end positions of the storyboard.

6. In the **Publish to Video** dialog, click the folder 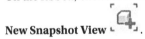 icon and specify the project folder as the **file location**.

7. Set **File Format** to **WMV File (*.wmv)**.

8. Select the **Reverse** option to reverse the animation.

9. Leave the other default settings and click **OK**; the **Publish Video Progress** dialog appears.

    A message box appears that the video has been published.

10. Click **OK** in the message box.

Taking a Snapshot of the Explosion

Follow these steps:

1. Click and drag the play marker on the timeline to 15 seconds.

    You can capture the snapshot of the current position of the assembly using the **New Snapshot View** tool.

2. On the ribbon, click the **Presentation** tab, go to the **Workshop** panel, and click

    New Snapshot View.

    The snapshot appears in the Snapshot Views window. Notice that marker on the snapshot. It indicates that the snapshot is dependent on the storyboard.

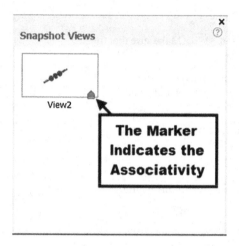

For example, if you make changes to the assembly at the position of the marker where the snapshot was taken, the Update View symbol appears on the snapshot view. You need to click the Update View symbol to update the snapshot.

3. Click **Save** ![icon] on the **Quick Access Toolbar**; the **Save As** dialog appears.

4. Type **Oldham_coupling** in the **File name**box.

5. Go to the project folder.

6. Click **Save** to save the file.

7. Click **OK**.

8. Click **File ➤ Close**.

CHAPTER 4

■ ■ ■

Creating Drawings

In this chapter, you will generate 2D drawings of the parts and assemblies.
In this chapter, you will do the following:

- Insert standard views of a part model
- Create centerlines and centermarks
- Retrieve model dimensions
- Add dimensions and annotations
- Create custom sheet formats and templates
- Insert an exploded view of an assembly into the drawing
- Insert a bill of materials of the assembly into the drawing
- Apply balloons to the assembly

Tutorial 1

In this tutorial, you will create the drawing of the Flange.ipt file created in Chapter 2.

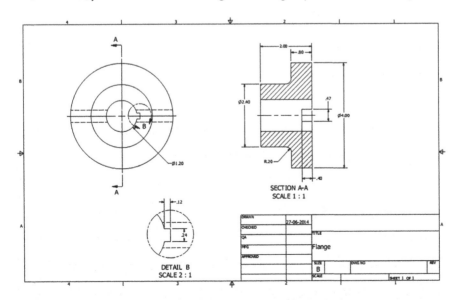

© T. Kishore 2017
T. Kishore, *Learn Autodesk Inventor 2018 Basics*, https://doi.org/10.1007/978-1-4842-3225-5_4

Starting a New Drawing File

Follow these steps:

1. To start a new drawing, click the **Drawing** icon on the Home screen.

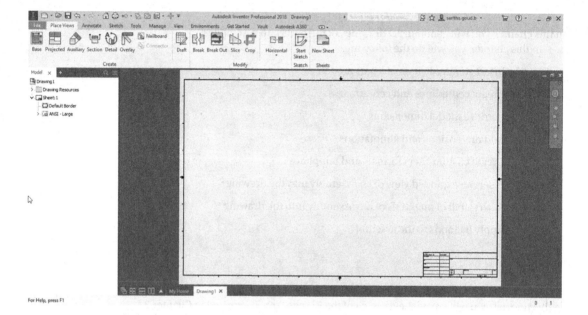

Editing the Drawing Sheet

Follow these steps:

1. To edit the drawing sheet, right-click **Sheet:1** in the **Browser window** and select **Edit Sheet** from the shortcut menu.

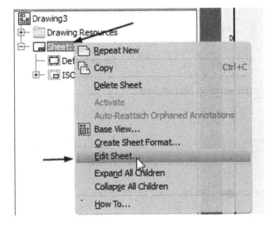

2. In the **Edit Sheet** dialog, set **Size** to **B**.

3. Click **OK**.

   The drawing views in this tutorial are created in the third-angle projection. If you want to change the type of projection, then continue with the steps.

4. Click **Manage ➤ Styles and Standards ➤ Styles Editor** on the ribbon.

5. In the **Style and Standard Editor** dialog, specify the settings shown here:

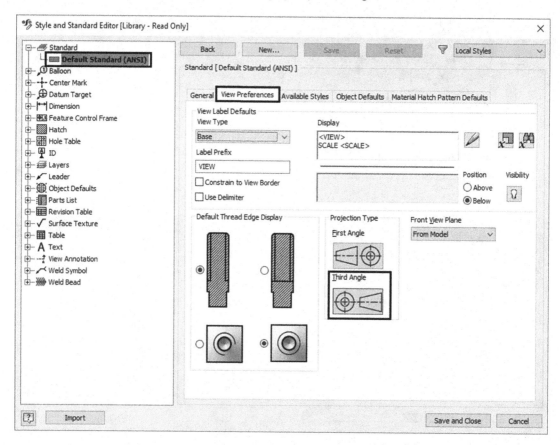

6. Click **Save and Close** .

Generating the Base View

Follow these steps:

1. To generate the base view, click **Place Views ➤ Create ➤ Base** on the ribbon.

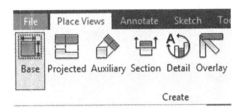

2. In the **Drawing View** dialog, click **Open Existing File** .

3. In the **Open** dialog, browse to the project folder.

4. Set **Files of type** to **Inventor Files (*.ipt, *.iam, *.ipn)** and then double-click **Flange.ipt**.

5. Set **Style** to **Hidden Line** .

6. Set **Scale** to **1:1**.

7. Click the preview, drag, and place it at the left side of the drawing sheet, as shown here.

8. Click **OK** in the dialog.

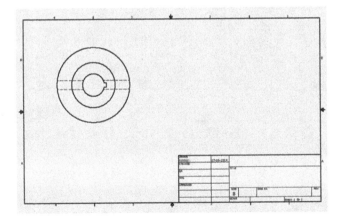

Generating the Section View

Follow these steps:

1. To create the section view, click **Place Views** ➤ **Create** ➤ **Section** on the ribbon.

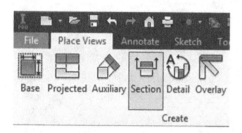

2. Select the base view.

3. Place the cursor on the top quadrant point of the circular edge, as shown here:

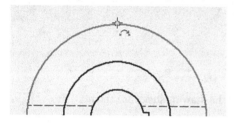

4. Move the pointer upward and notice the dotted line.

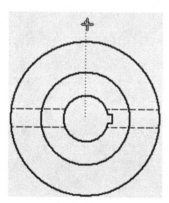

5. Click the dotted line and move the cursor vertically downward.

6. Click outside the bottom portion of the view, as shown here:

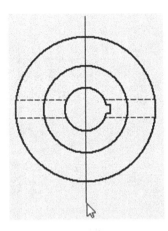

7. Right-click and select **Continue**.

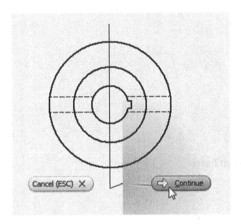

8. Move the cursor toward the right and click to place the section view.

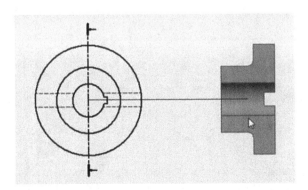

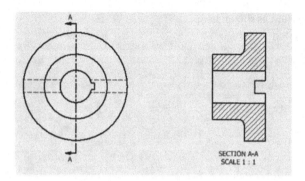

Creating the Detailed View

Now you have to create the detailed view of the Center Marks, which is displayed in the front view.

    1.   To create the detailed view, click **Place Views ➤ Create ➤ Detail** on the ribbon.

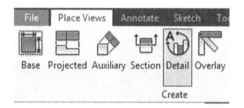

    2.   Select the base view.

    3.   In the **Detail View** dialog, specify the settings as shown here:

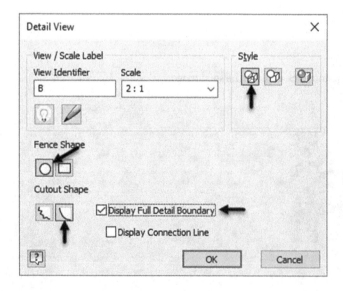

4. Specify the center point and boundary point of the detail view, as shown here:

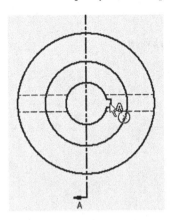

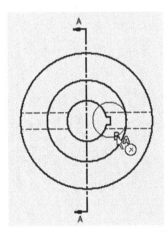

5. Place the detail view below the base view.

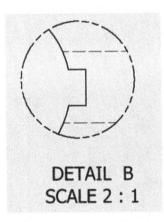

Creating Centermarks and Centerlines

Follow these steps:

1. To create a center mark, click **Annotate ➤ Symbols ➤ Center Mark** on the ribbon.

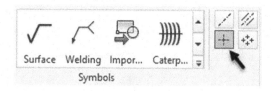

2. Click the outer circle of the base view.

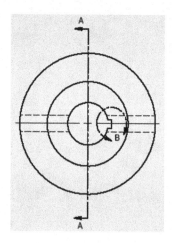

3. To create a centerline, click **Annotate ➤ Symbols ➤ Centerline Bisector** on the ribbon.

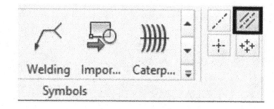

4. Click the inner horizontal edges of the section view.

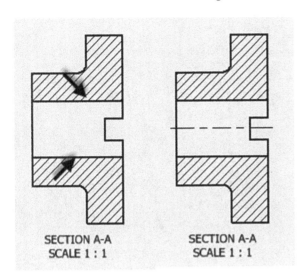

Retrieving Dimensions

Now you will retrieve the dimensions that were applied to the model while creating it.

1. To retrieve dimensions, click **Annotate ➤ Retrieve ➤ Retrieve Model Annotations** on the ribbon.

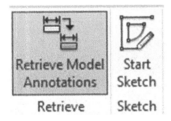

The **Retrieve Model Annotation** dialog appears.

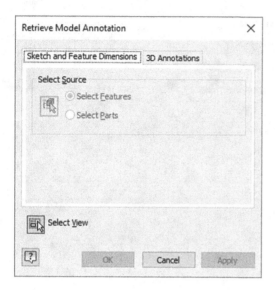

2. Select the section view from the drawing sheet.

Now, you must select the dimensions to retrieve.

3. Drag a window on the section view to select all the dimensions.

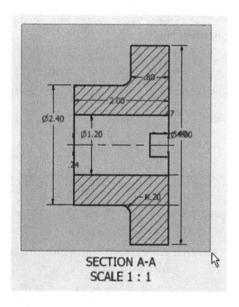

4. Click **Select Features** under the **Select Source** group.

5. Click **OK** to retrieve feature dimensions.

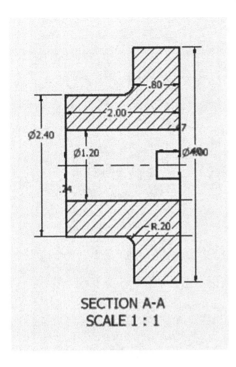

SECTION A-A
SCALE 1 : 1

6. Click **Annotate ➤ Dimension ➤ Arrange** on the ribbon.

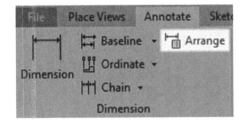

7. Drag a selection box and select all the dimensions of the section view.

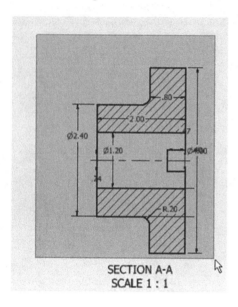

8. Right-click and select **OK**.

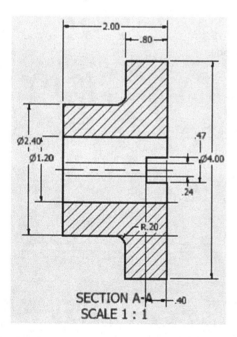

9. Select the unwanted dimensions and press Delete.

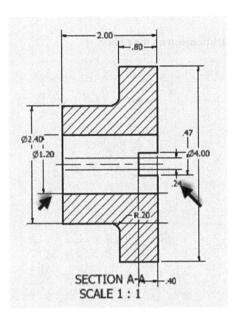

10. Click and drag the dimensions to arrange them properly.

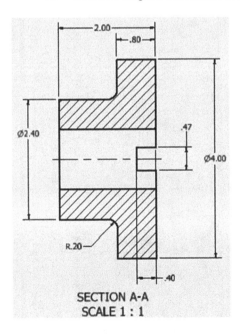

Adding Dimensions

Follow these steps:

1. To add dimensions, click **Annotate ➤ Dimension ➤ Dimension** on the ribbon.

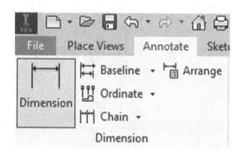

2. Select the center hole on the base view.

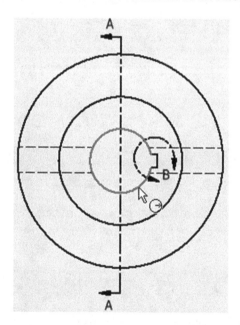

3.   Right-click and select **Dimension Type ➤ Diameter**.

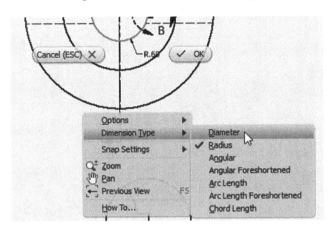

4.   Place the dimension, as shown here. The **Edit Dimension** dialog appears.

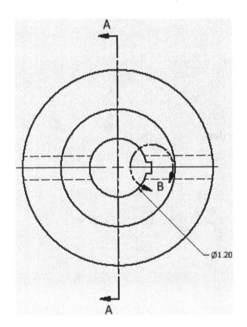

5.   Click **OK**.

6. Create the dimensions on the detail view, as shown here:

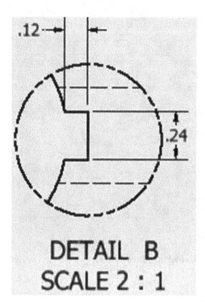

Populating the Title Block

Follow these steps:

1. To populate the title block, click **File ➤ iProperties**.

2. In the **Flange iProperties** dialog, click the tabs one by one and type data in the respective fields.

3. Click **Apply** and **Close**.

Saving the Drawing

Follow these steps:

1. Click **Save** on the **Quick Access Toolbar**; the **Save As** dialog appears.

2. Type **Flange** in the **File name** box.

3. Go to the project folder.

4. Click **Save** to save the file.

5. Click **File ➤ Close**.

Tutorial 2

In this tutorial, you will create a custom template and then use it to create a new drawing.

Creating a New Sheet Format

Follow these steps:

1. On the ribbon, click **Get Started ➤ Launch ➤ New**.

2. In the **Create New File** dialog, click the **Standard.idw** icon.

Standard.idw

3. Click **Create** to start a new drawing file.

4. To edit the drawing sheet, right-click **Sheet:1** in the **Browser window** and select **Edit Sheet** from the shortcut menu.

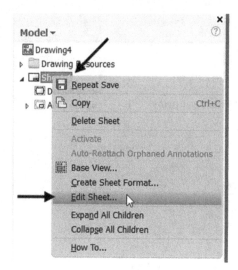

5. In the **Edit Sheet** dialog, set **Size** to **B**.

Under the **Orientation** section, you can change the orientation of the title block as well as the sheet orientation.

6. Click **OK**.

7. In the **Browser window**, expand the **Drawing Resources ➤ Sheet Formats** folder to view different sheet formats available. Now, you will add a new sheet format to this folder.

8. Right-click the **Borders** folder and select **Define New Border**.

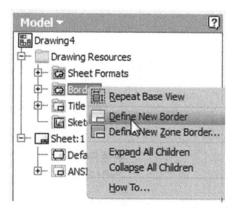

Now, you can create a new border using the sketch tools available on the Sketch tab.

9. Click **Finish Sketch** on the **Sketch** tab of the ribbon.

10. In the **Border** dialog, click **Discard**.

11. In the Browser window, right-click the **Borders** folder and select **Define New Zone Border**.

12. In the **Default Drawing Border Parameters** dialog, type **4** in the **Vertical Zones** box and click **OK**.

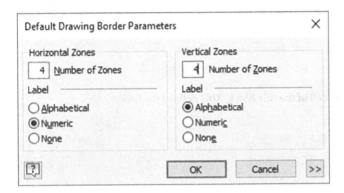

13. Click **Finish Sketch** on the **Sketch** tab of the ribbon.

14. In the **Border** dialog, type **4-Zone Border** and click **Save**.

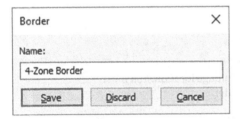

15. Expand the **Title Blocks** folder and right-click **ANSI-Large**.

16. Select **Edit** from the shortcut menu.

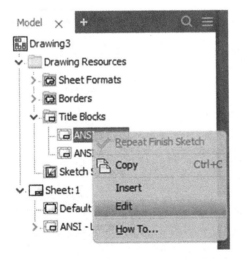

17. On the **Sketch** tab of the ribbon, click **Insert ➤ Image**.

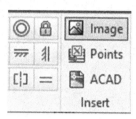

18. Draw a rectangle in the **Company** cell of the title block. This defines the image size and location.

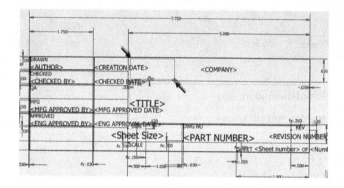

19. Go to the location of your company logo or any other image location. You must ensure that the image is located inside the project folder.

20. Select the image file and click **Open**. This will insert the image into the title block.

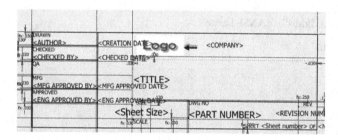

21. Click **Finish Sketch** on the ribbon.

22. Click **Save As** in the **Save Edits** dialog.

23. Type **ANSI-Logo** in the **Title Block** dialog.

24. Click **Save**.

25. In the Browser window, expand **Sheet:1** and right-click **Default Border**.

26. Select **Delete** from the shortcut menu.

27. Expand the **Borders** folder and right-click **4-Zone Border**.

28. Select **Insert** from the shortcut menu.

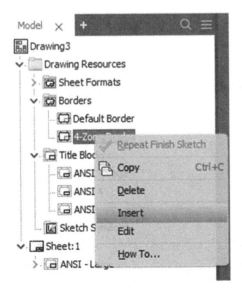

29. Click **OK** in the **Edit Drawing Border Parameters** dialog.

30. Expand **Sheet:1** and right-click **ANSI-Large**.

31. Select **Delete** from the shortcut menu.

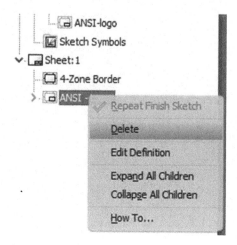

32. Expand the **Title Blocks** folder and right-click **ANSI-Logo**.

33. Select **Insert** from the shortcut menu to insert the title block.

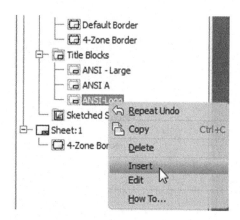

112

34. Right-click **Sheet:1** and select **Create Sheet Format**.

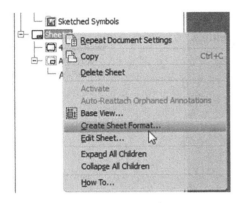

35. Type **Custom Format** in the **Create Sheet Format** dialog and then click **OK**.

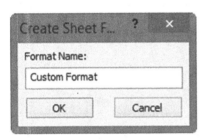

You will notice that the new sheet format is listed in the **Sheet Formats** folder.

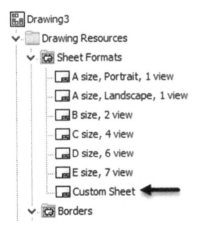

Creating a Custom Template

Follow these steps:

1. On the ribbon, click **Tools** ➤ **Options** ➤ **Document Settings**.

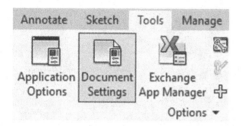

In **Document Settings** dialog, you can define the standard, sheet color, drawing view settings, and sketch settings.

2. Leave the default settings in this dialog and click **Close**.

3. In the Browser window, expand the **Sheet Formats** folder and double-click **Custom Format**.

4. Right-click **Sheet: 2** and select **Delete Sheet** from the shortcut menu.

5. Click **OK**.

6. On the ribbon, click **Manage ➤ Styles and Standards ➤ Styles Editor**.

7. In the **Style and Standard Editor** dialog, select **Dimension ➤ Default (ANSI)**.

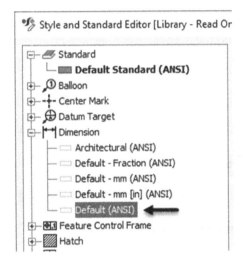

8. Click the **New** button located at the top of the dialog.

9. Type **Custom Standard** in the **New Local Style** dialog and then click **OK**.

10. Click the **Units** tab and set **Precision** to **3.123**.

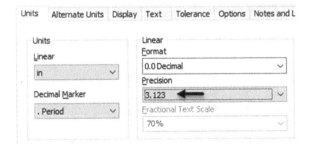

11. Click **Save and Close**.

12. On the **File menu**, click **Save As ➤ Save Copy As Template**. This will take you to the templates folder on your drive.

13. Type **Custom Template** in the **File name** box.

14. Click **Save**.

15. Close the drawing file without saving it.

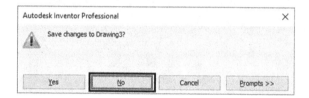

Starting a Drawing Using the Custom Template

Follow these steps:

1. On the ribbon, click **Get Started ➤ Launch ➤ New**.

2. In the **Create New File** dialog, click the **Custom Template.idw** icon.

Custom
Template.idw

3. Click **Create** and **OK** to start a new drawing file.

Generating the Drawing Views

Follow these steps:

1. To generate views, click **Place Views ➤ Create ➤ Base** on the ribbon.

2. In the **Drawing View** dialog, click **Open Existing File**.

3. Go to the project folder and double-click **Disc.ipt**.

4. Select **Front** from the ViewCube displayed on the sheet.

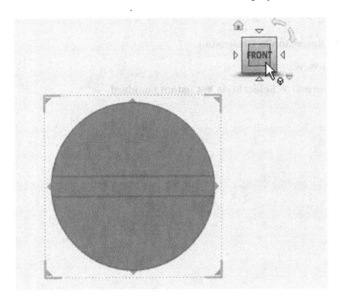

5. Set **Scale** to **1:1**.

6. Click and drag the view to the top center of the drawing sheet.

7. Move the cursor downward and click to place the projected view.

8. Right-click and select **OK**.

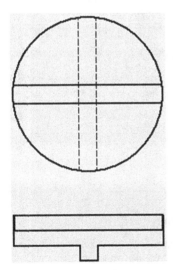

Adding Dimensions

Follow these steps:

1. On the ribbon, click **Annotate ➤ Dimension ➤ Dimension**.

2. Select the circular edge on the base view.

3. On the ribbon, click **Annotate ➤ Format ➤ Select Style ➤ Custom Standard**.

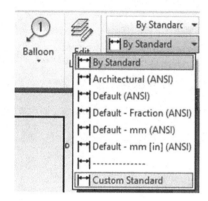

4. Right-click and select **Dimension Type ➤ Diameter**.

5. Click to place the dimension.

6. Click **OK**.

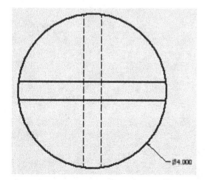

7. Select the horizontal edges on the base view.

8. Move the pointer toward the right and click to place the dimension.

9. Click **OK** in the **Edit Dimension** dialog.

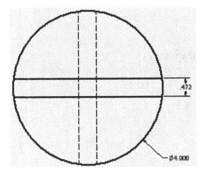

10. Add other dimensions to the drawing.

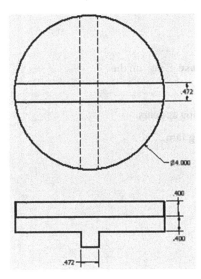

11. Right-click and select **OK** to deactivate the **Dimension** tool.

12. Save and close the drawing file.

Tutorial 3

In this tutorial, you will create the drawing of the Oldham coupling assembly created in the previous chapter.

Creating a New Drawing File

Follow these steps:

1. On the ribbon, click **Get Started ➤ Launch ➤ New**.

2. In the **Create New File** dialog, click the **Custom Template.idw** icon.

Custom
Template.idw

3. Click **Create** and **OK** to start a new drawing file.

Generating the Base View

Follow these steps:

1. To generate the base view, click **Place Views ➤ Create ➤ Base** on the ribbon; the **Drawing View** dialog appears.

2. Click **Open Existing File** in this dialog; the **Open** dialog appears.

3. Go to the project folder and double-click **Oldham_Coupling.iam**.

4. Click the **Home** icon located above the ViewCube.

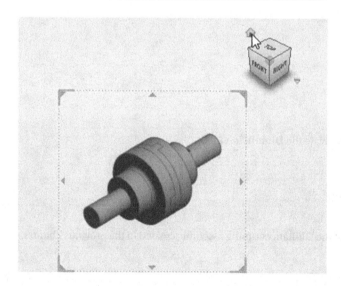

5. Set **Scale** to **1/2**.

6. Click and drag the view to the top-left corner.

7. Right-click and select **OK**.

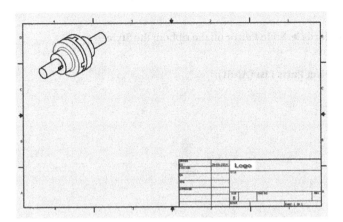

Generating the Exploded View

Follow these steps:

1. To generate the base view, click **Place Views ➤ Create ➤ Base** on the ribbon; the Drawing View dialog appears.

2. Click **Open Existing File** in this dialog; the **Open** dialog appears.

3. Go to the project folder and double-click **Oldham_Coupling.ipn**.

4. Click the **Home** icon located above the ViewCube.

5. Set **Scale** to **1/2**.

6. Click and drag the view to the center of the drawing sheet.

7. Right-click and select **OK**.

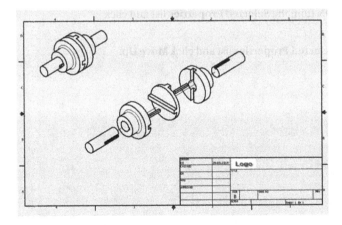

Configuring the Parts List Settings

Follow these steps:

1. Click **Manage ➤ Styles and Standards ➤ Style Editor** on the ribbon; the **Style and Standard Editor** dialog appears.

2. Expand the **Parts List** node and select **Parts List (ANSI)**.

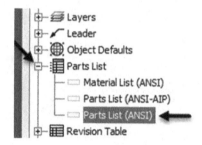

3. Click the **Column Chooser** button under the **Default Columns Settings** group; the **Parts List Column Chooser** dialog appears.

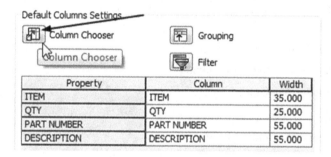

4. In this dialog, select **DESCRIPTION** from the **Selected Properties** list and click the **Remove** button.

5. Select **PART NUMBER** from the **Selected Properties** list and click **Move Up**.

6. Click **OK**.

7. Click **Save** and **Close**.

Creating the Parts List

Follow these steps:

1. To create a parts list, click **Annotate ➤ Table ➤ Parts List** on the ribbon; the **Parts List** dialog appears.

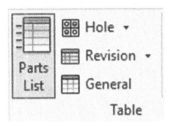

2. Select the exploded view from the drawing sheet.

3. Select **Parts Only** from the **BOM View** drop-down under the **BOM Settings and Properties** group.

4. Click **OK** twice.

5. Place the part list above the title block.

PARTS LIST		
ITEM	PART NUMBER	QTY
1	Disc	1
2	Flange	2
3	Shaft	2
4	Key	2

Creating Balloons

Follow these steps:

1. To create balloons, click **Annotate ➤ Table ➤ Balloon ➤ Auto Balloon** on the ribbon; the **Auto-Balloon** dialog appears.

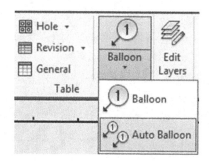

2. Select the exploded view from the drawing sheet.

3. Select all the parts in the exploded view.

4. Select **Horizontal** from the **Placement** group.

5. Click the **Select Placement** button in the **Placement** group.

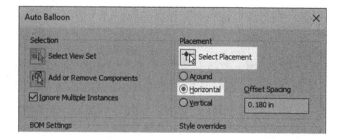

6. Click above the exploded view.

7. Click **OK** to place the balloons.

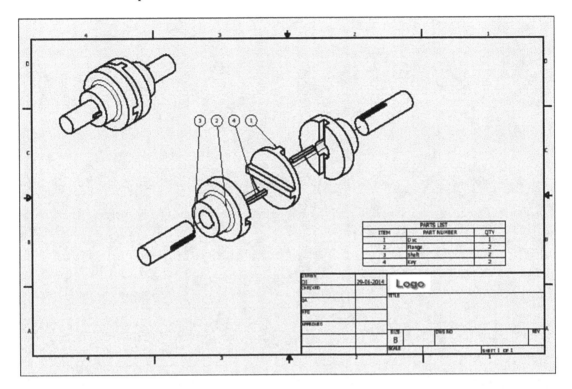

Saving the Drawing

Follow these steps:

1. Click **Save** on the **Quick Access Toolbar**; the **Save As** dialog appears.

2. Type **Oldham_Coupling** in the **File name** box.

3. Go to the project folder.

4. Click **Save** to save the file.

5. Click **OK**.

6. Click **File ➤ Close**.

CHAPTER 5

■ ■ ■

Additional Modeling Tools

In this chapter, you will create models using additional modeling tools. You will learn to do the following:

- Create slots
- Create circular patterns
- Create holes
- Create chamfers
- Create shells
- Create rib features
- Create coils
- Create a loft feature
- Create an emboss feature
- Create a thread
- Create a sweep feature
- Create a grill feature
- Create a replace faces
- Create a face fillet
- Create a variable fillet
- Create a boss feature
- Create a lip feature

© T. Kishore 2017
T. Kishore, *Learn Autodesk Inventor 2018 Basics*, https://doi.org/10.1007/978-1-4842-3225-5_5

Tutorial 1

In this tutorial, you will create the model shown here:

Creating the First Feature

Follow these steps:

1. Create a new project with the name **Autodesk Inventor 2018 Basics Tutorial** (see Chapter 2's Tutorial 1 to learn how to create a new project).

2. Open a new Inventor part file using the **Standard.ipt** template (see Chapter 2's Tutorial 3).

3. Click the **Start 2D Sketch** button on the ribbon and select XY Plane.

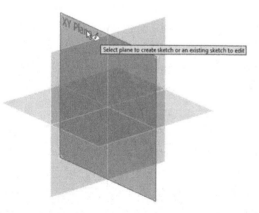

4. Click the **Circle Center Point** 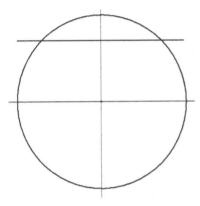 button and draw a circle (see Chapter 2's Tutorial 1).

5. Click the **Line** button.

6. Specify a point at the top left outside the circle.

7. Move the pointer horizontally and notice the horizontal constraint symbol.

8. Click outside the circle. Press Esc to deactivate the **Line** tool.

9. Click the **Trim** button on the **Modify** panel.

10. Click the portions of the sketch you want to trim, as shown here:

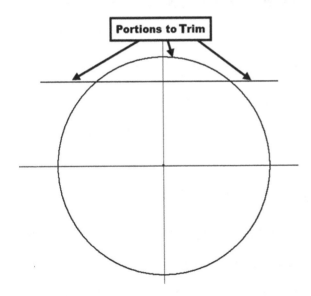

11. Apply dimensions to the sketch (radius = 0.63, vertical length = 1.102). To apply the vertical length dimension, activate the **Dimension** tool and select the horizontal line. Move the pointer downward and place the cursor on the bottom quadrant point of the arc. Click when the symbol appears. Move the pointer toward the left and click to place the dimension.

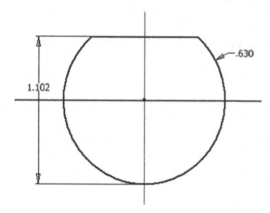

12. Click **Rectangle ➤ Slot Center Point Arc** on the **Create** panel.

13. Select the origin as the center point.

14. Move the cursor outside and click in the first quadrant of the circle to specify the start point of the slot arc.

15. Move the cursor and click in the fourth quadrant of the circle to specify the end point of the slot arc.

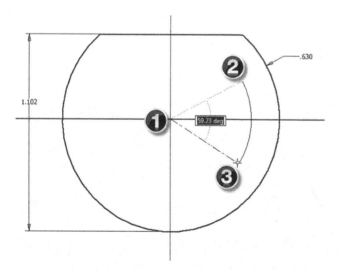

16. Move the cursor outward from the arc and click.

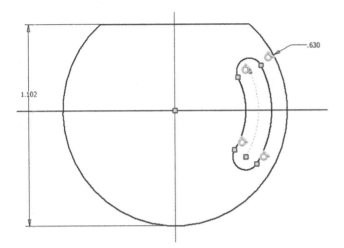

17. Click the **Dimension** button on the **Constrain** panel.

18. Select the start point of the slot arc.

19. Select the center point of the slot arc.

20. Select the end point of the slot arc.

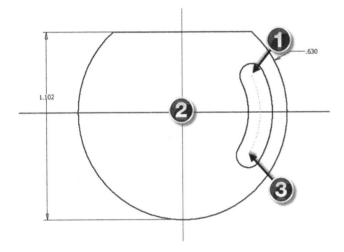

21. Place the angular dimension of the slot; the **Edit Dimension** box appears.

22. Enter **30** in the **Edit Dimension** box and click the green check.

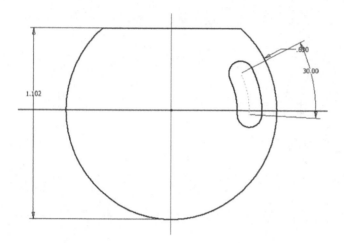

23. Click the **Construction** button on the **Format** panel.

24. Click the **Line** button on the **Create** panel.

25. Draw a horizontal line passing through the origin.

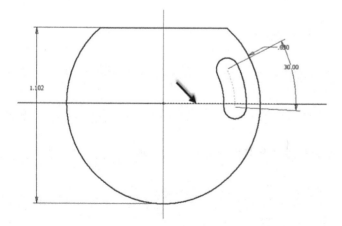

26.   Click the **Symmetric** 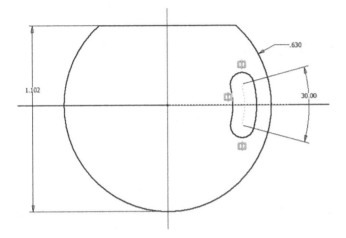 button on the **Constrain** panel.

27.   Select the end caps of the slot.

28.   Select the construction line; the slot is made symmetric about the construction line.

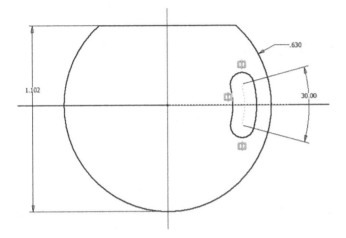

29.   Apply other dimensions to the slot.

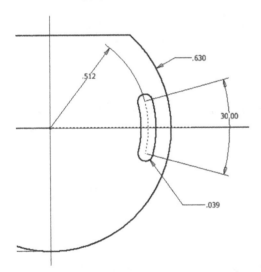

30. Click the **Circular Pattern** button on the **Pattern** panel; the **Circular Pattern** dialog appears.

31. Select all the elements of the slot by creating a selection window.

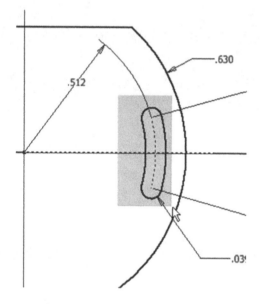

32. Click the cursor button located on the right side of the dialog.

33. Select the origin point of the sketch axis point.

34. Enter **4** in the **Count** box and **180** in the **Angle** box.

35. Click the **Flip** button.

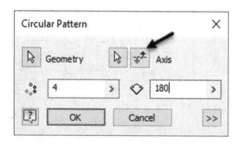

The preview of the circular pattern appears.

36.   Click **OK** to create the circular pattern.

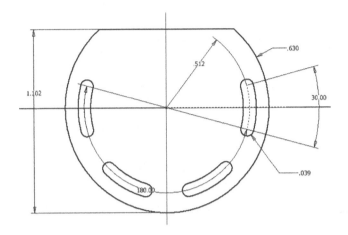

37.   Click the **Finish Sketch** button on the ribbon.

38.   Extrude the sketch up to a distance of 0.236 (see Chapter 2's Tutorial 1).

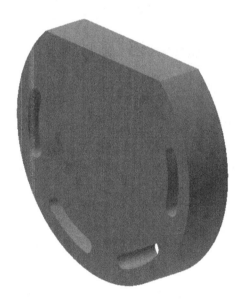

Adding the Second Feature

Follow these steps:

1. Create a sketch on the back face of the model (use the Orbit tool available on the Navigation Bar to rotate the model).

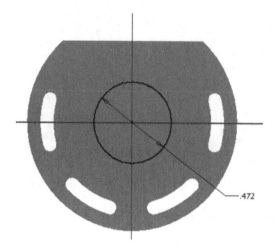

2. Extrude the sketch up to a distance of 0.078.

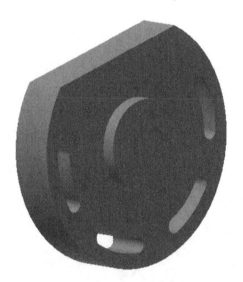

Creating a Counterbore Hole

In this section, you will create a counterbore hole concentric to the cylindrical face.

1. Click the **Hole** button on the **Modify** panel; the **Hole** dialog appears.

2. Set the parameters in the **Hole** dialog, as shown here:

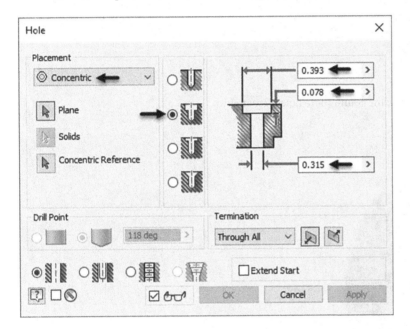

3. Click the front face of the model; the preview of the hole appears.

Now, you need to specify the concentric reference.

4. Select the cylindrical face of the model; the hole is made concentric to the model.

5. Click **OK** in the dialog; the counterbore hole is created.

Creating a Threaded Hole

In this section, you will create a hole using a sketch point.

1. Click the **Start 2D Sketch** 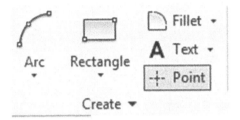 button and select the front face of the model.

2. Click the **Point** button on the **Create** panel.

3. Place the point on the front face of the model.

4. Click the **Horizontal** button on the **Constrain** panel.

5. Select the point and sketch origin; the point becomes horizontal to the origin.

6. Create a horizontal dimension of 0.354 between the point and origin.

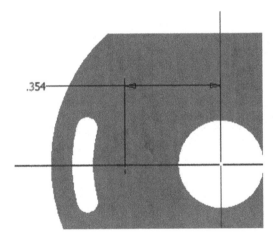

7. Click **Finish Sketch**.

8. Click the **Hole** button on the **Modify** panel; the **Hole** dialog appears.

9. In the **Hole** dialog, set **Placement** to **From Sketch**.

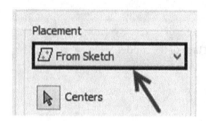

10. Select the **Counterbore** option.

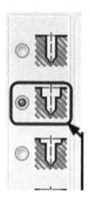

11. Set the **counterbore diameter** to 0.118.

12. Set the **counterbore depth** to 0.039.

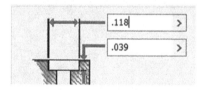

13. Select the **Tapped Hole** option.

14. Set **Thread Type** to **ANSI Unified Screw Threads**.

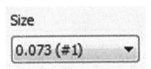

15. Set **Size** to **0.073**.

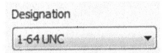

16. Set **Designation** to **1-64 UNC**.

17. Select the **Full Depth** option.

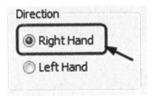

18. Set **Direction** to **Right Hand**.

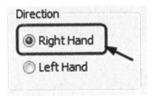

19. Click **OK** to create the hole.

Creating a Circular Pattern

Follow these steps:

1. Click the **Circular Pattern** button on the **Pattern** panel; the **Circular Pattern** dialog appears.

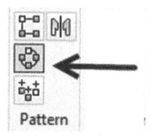

2. Select the threaded hole created in the previous section.

3. Click the **Rotation Axis** button in the dialog.

4. Select the outer cylindrical face of the model.

5. Enter **6** in the **Occurrence** box and **360** in the **Angle** box.

6. Click **OK** to create the circular pattern.

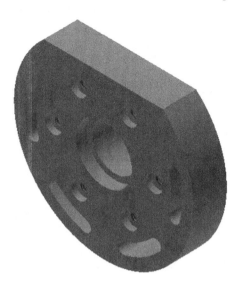

Creating Chamfers

Follow these steps:

1. Click the **Chamfer** ⬦ button on the **Modify** panel.
2. Click the **Distance and Angle** button in the dialog.

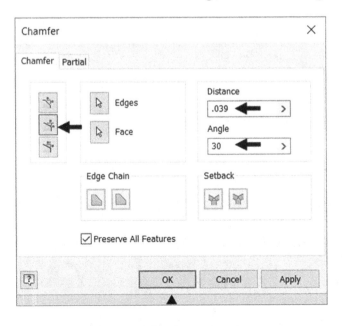

3. Select the cylindrical face of the counterbore hole located at the center.

Face Selected

4. Select the circular edge of the counterbore hole.

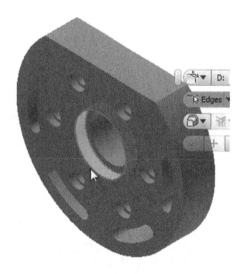

5. Enter **0.039** in the **Distance** box and **30** in the **Angle** box.

6. Click **OK** to create the chamfer.

7. Save the model and close it.

Tutorial 2

In this tutorial, you will create the model shown here:

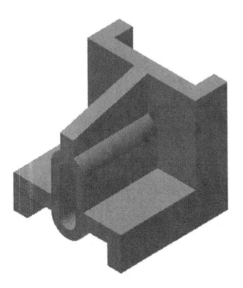

Creating the First Feature

Follow these steps:

1. Open a new Inventor part file using the **Standard.ipt** template (see Chapter 2's Tutorial 3).

2. On the ribbon, click **3D Model ➤ Sketch ➤ Start 2D Sketch** .

3. Select the YZ plane.

4. Draw an L-shaped sketch using the **Line** tool and dimension it to be 1.575, as shown here:

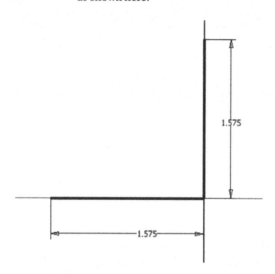

5. Click the **Offset** button on the **Modify** panel.

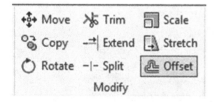

6. Select the sketch and then specify the offset position at a random distance.

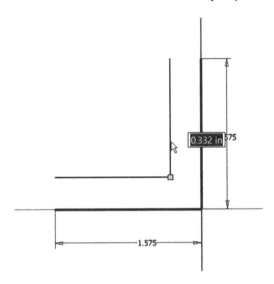

7. Click the **Line** button and draw lines closing the offset sketch.

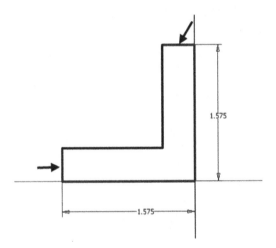

8. Add the offset dimension to the sketch.

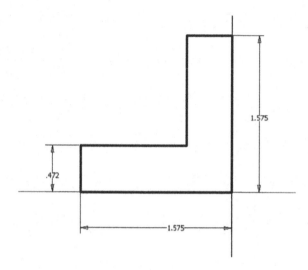

9. Click **Finish Sketch**.

10. Click **3D Model ➤ Create ➤ Extrude** 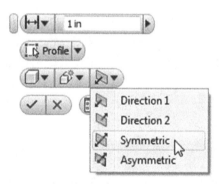 on the ribbon.

11. Select the **Symmetric** option from the mini-toolbar.

12. Set **Distance** to 1.575.

13.　Click **OK** to create the first feature

Creating the Shell Feature

You can create a shell feature by removing a face of the model and applying thickness to other faces.

1.　Click **3D Model ➤ Modify ➤ Shell** ⬛ on the ribbon; the **Shell** dialog
appears.

2.　Set **Thickness** to 0.197.

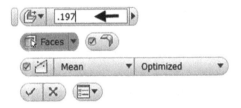

Now, you need to select the faces to remove.

3. Select the top face and the back face of the model.

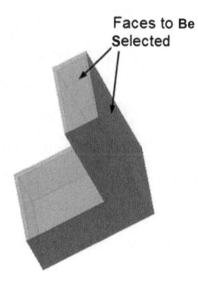

4. Select the front face and the bottom face of the model.

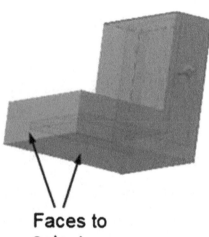

5.   Click **OK** to shell the model.

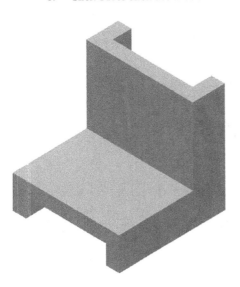

Creating the Third Feature

Follow these steps:

1.   Click **3D Model ➤ Sketch ➤ Start 2D Sketch** on the ribbon.

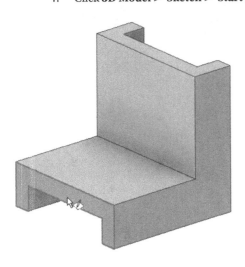

2.   Select the front face of the model.

3. Click **Sketch ➤ Create**, select the **Rectangle** drop-down, and select **Slot Center to Center** on the ribbon.

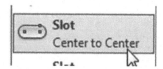

4. Draw a slot by selecting the first, second, and third points. Make sure that the second point is coincident with the lower horizontal edge.

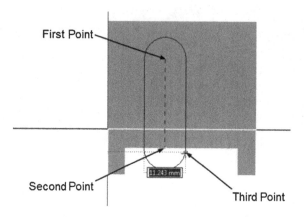

5. Apply dimensions to the slot.

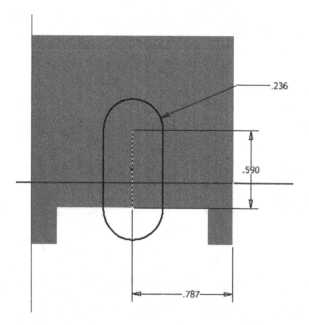

6. Click **Finish Sketch**.

7. Click **3D Model ➤ Create ➤ Extrude** on the ribbon.

8. Select the sketch.

9. Select the **To** option from the **Extents** drop-down.

10. Select the back face of the model.

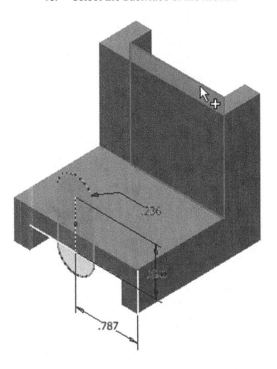

11. Click the **Join** button in the dialog.

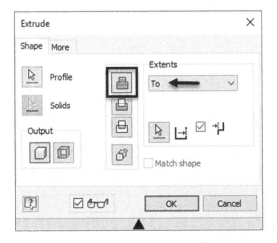

12. Click **OK** to create the feature.

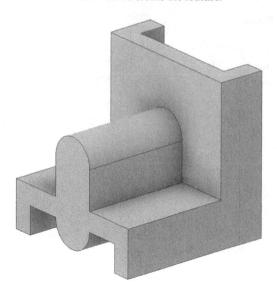

Creating a Cut Feature

Follow these steps:

1. Create the sketch on the front face of the model, as shown here:

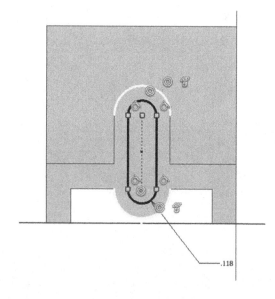

2. Finish the sketch.

3. Click **3D Model ➤ Create ➤ Extrude** on the ribbon.

4. Select the sketch.

5. Select the **All** option from the **Extents** drop-down.

6. Click the **Cut** button in the dialog.

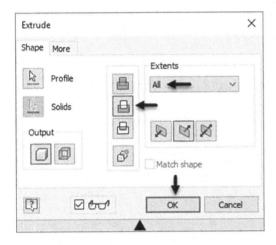

7. Click **OK** to create the cut feature

Creating the Rib Feature

In this section, you will create a rib feature at the middle of the model. To do this, you must create a midplane.

1. To create a midplane, click **3D Model ➤ Work Features ➤ Plane ➤ Midplane between Two Planes** on the ribbon.

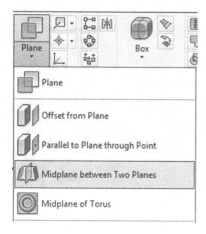

2. Select the right face of the model.

3. Select the left face of the model; the midplane is created.

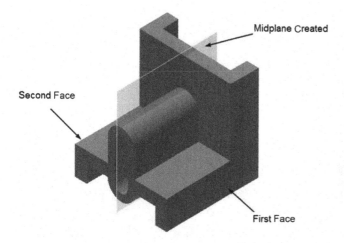

4. Click **3D Model ➤ Sketch ➤ Start 2D Sketch** on the ribbon.

5. Select the midplane.

6. Click the **Slice Graphics** button at the bottom of the window.

7. Click **Sketch ➤ Create ➤ Project Geometry ➤ Project Cut Edges** on the ribbon; the edges cut by the sketch plane are projected.

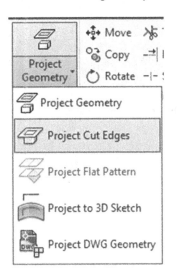

8. Draw the sketch, as shown here:

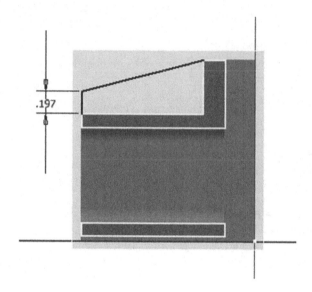

9. Finish the sketch.

10. Click **3D Model ➤ Create ➤ Rib** on the ribbon; the **Rib** dialog appears.

11. Select the sketch.

12. Click the **Parallel to Sketch Plane** button in the dialog.

13. Click the **Direction 1** button.

14. Set **Thickness** to 0.197.

15. Click the **To Next** button.

16. Click the **Symmetric** button below the **Thickness** box.

17. Click **OK** to create the rib feature.

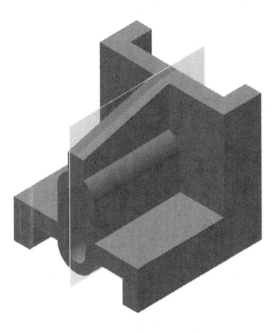

18. To hide the midplane, select it and right-click.

19. Click **Visibility** on the marking menu; the plane will be hidden.

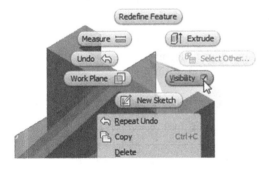

20. Save the model and close it.

Tutorial 3

In this tutorial, you will create a helical spring using the **Coil** tool.

Creating the Coil

Follow these steps:

1. Open a new Inventor file using the **Standard.ipt** template (see Chapter 2's Tutorial 3).

2. Click the **Start 2D Sketch** button on the ribbon and select XY Plane.

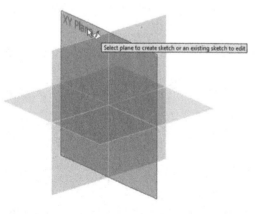

3. Click the **Circle Center Point** button.

4. Select a point on the left portion of the horizontal axis.

5. Move the cursor outward and click to create a circle.

6. On the ribbon, click **Sketch ➤ Format ➤ Centerline** .

7. Click the **Line** button.

8. Select the origin point of the sketch, move the cursor upward, and click to create a centerline.

9. On the ribbon, click **Sketch ➤ Constrain ➤ Dimension**.

10. Add the diameter dimension to the circle, as shown next.

11. Select the centerline and circle.

12. Move the pointer downward and click.

13. Type **1.575** and press Enter.

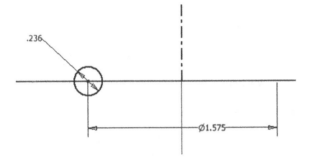

14. Finish the sketch.

15. To create a coil, click **3D Model ➤ Create ➤ Coil** on the ribbon; the **Coil** dialog appears.

    In addition, the profile is automatically selected. Now, you need to select the axis of the coil.

16. Select the centerline as the axis. Click the **Reverse Axis** button in the **Coil** dialog, if the coil preview is downward.

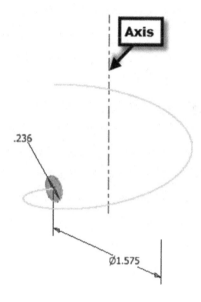

17. Click the **Coil Size** tab in the dialog.

18. On the **Coil Size** tab, specify the settings, as shown here:

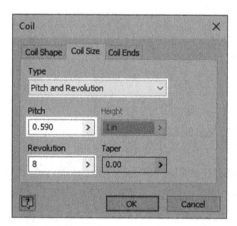

19. Click the **Coil Ends** tab in the dialog.

20. Specify the settings on the **Coil Ends** tab, as shown here:

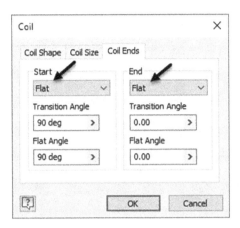

21. Click **OK** to create the coil.

22. Save the model as **Coil.ipt** and close the file.

Tutorial 4

In this tutorial, you create a shampoo bottle using the **Loft, Extrude,** and **Coil** tools.

Creating the First Section and Rails

To create a swept feature, you need to create sections and guide curves.

1. Open a new part file (see Chapter 2's Tutorial 3).

2. Click **3D Model ➤ Sketch ➤ Start 2D Sketch** on the ribbon.

3. Select the XZ plane.

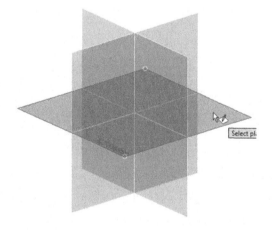

4. Click **Sketch ➤ Create ➤ Circle ➤ Ellipse** 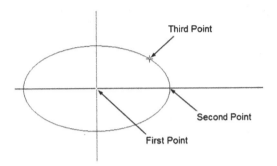 on the ribbon.

5. Draw the ellipse by selecting the points, as shown here:

Third Point

Second Point

First Point

6. On the ribbon, click **Sketch ➤ Constrain ➤ Dimension**.

7. Select the ellipse, move the cursor downward and click.

8. Type **1.968** and press Enter.

9. Select the ellipse, move the cursor toward the left, and click.

10. Type **0.984** and press Enter.

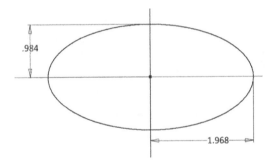

.984

1.968

11. Click **Finish Sketch**.

12. Click **3D Model ➤ Sketch ➤ Start 2D Sketch** on the ribbon.

13. Select the YZ plane from the graphics window.

14.   Click **Sketch ➤ Create ➤ Line ➤ Spline Interpolation** on the ribbon.

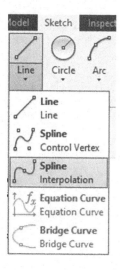

15.   Select a point on the horizontal axis of the sketch; a rubber band curve is attached to the cursor.

16.   Move the cursor up and specify the second point of the spline; a curve is attached to cursor.

17.   Move the cursor up and specify the third point.

18.   Likewise, specify the other points of the spline, as shown next.

19.   Right-click and select **Create**. The spline will be similar to the one shown here:

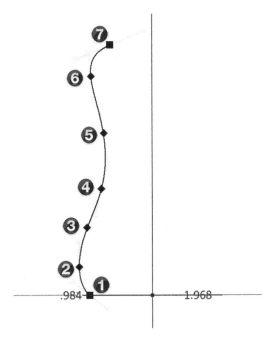

20.   Right-click and select **Create Line** from the marking menu.

21.   On the ribbon, click **Sketch ➤ Format ➤ Construction** .

22. Select the origin point of the sketch, move the pointer vertically upward, and click to create a vertical construction line.

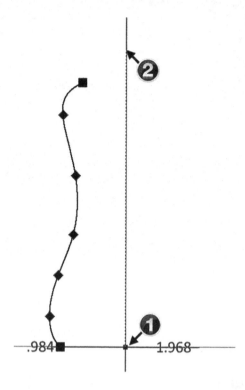

23. On the ribbon, click **Sketch ➤ Constrain ➤ Vertical** ⫴ .

24. Select the lower-end points of the vertical construction line and the spline.

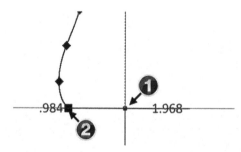

25. On the ribbon, click **Sketch ➤ Constrain ➤ Dimension**.

26. Select the lower end points of the construction line and spline.

27. Move the cursor downward and click.

28. Type **1.968** in the Edit Dimension box and press Enter.

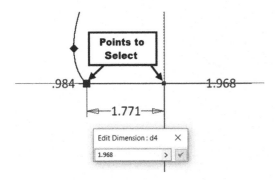

29. Select the second point of the spline and the construction line.

30. Move the cursor and click to place the dimension.

31. Type **2.362** in the **Edit Dimension** box and press Enter.

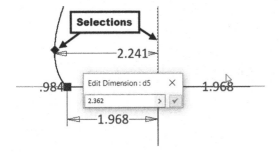

32.   Apply the other horizontal dimensions to the spline, as shown here:

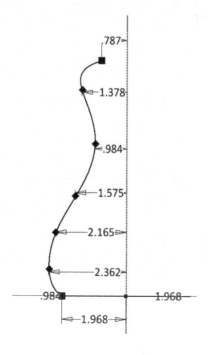

33.   Select the origin point of the sketch and the top-end point of the spline.

34.   Move the pointer toward the right and click.

35.   Type **8.858** in the **Edit Dimension** box and press Enter.

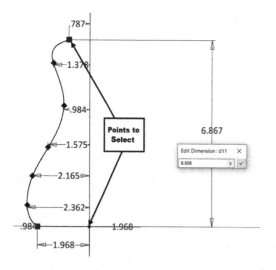

36. Likewise, create other dimensions, as shown here:

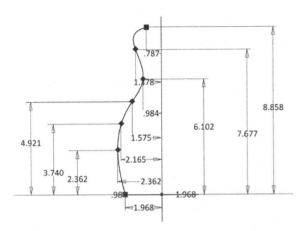

37. Click **Sketch ➤ Pattern ➤ Mirror** on the ribbon; the **Mirror** dialog appears.

38. Select the spline. Make sure you select the curve and not the points.

39. Click **Mirror Line** in the **Mirror** dialog and then select the construction line.

40. Click **Apply** and then click **Done**.

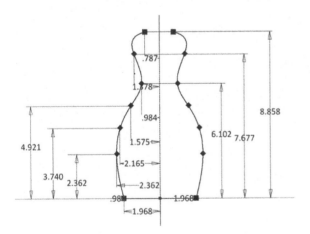

41. Click **Finish Sketch**.

Creating the Second Section

Follow these steps:

1. Click **3D Model ➤ Work Features ➤ Plane ➤ Offset from Plane** on the ribbon.

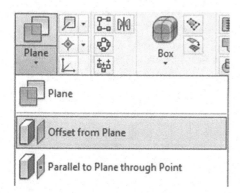

2. Select the XZ plane from the Browser window.

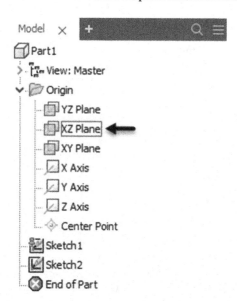

3. Enter **8.858** in the **Distance** box.

4. Click **OK** .

5. On the ribbon, click **3D Model ➤ Sketch ➤ Start 2D Sketch**.

6. Select the newly created datum plane.

7. Right-click and select **Center Point Circle** from the marking menu.

8. Select the origin point, move the cursor outside, and click to create a circle.

9. Right-click and select **OK**.

10. Right-click and select **General Dimension** from the marking menu.

11. Select the circle, move the cursor outward, and click.

12. Type **1.574** and press Enter.

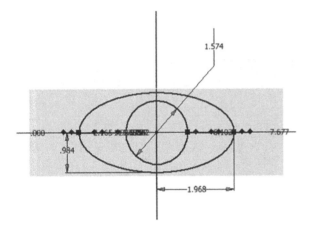

13. Click **Finish Sketch** on the ribbon.

Creating the Loft Feature

Follow these steps:

1. To create a loft feature, click **3D Model ➤ Create ➤ Loft** 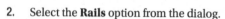 on the ribbon; the **Loft** dialog appears.

2. Select the **Rails** option from the dialog.

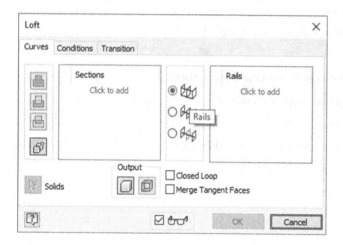

3. Click "**Click to add**" in the **Sections** group and select the circle.

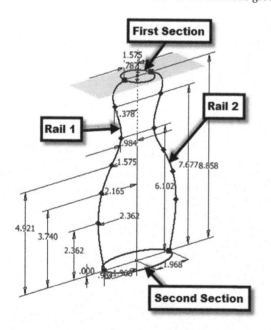

4. Select the ellipse.

5. Click "**Click to add**" in the **Rails** group.

6. Select the first rail.

7. Select the second rail.

8. Click **OK** to create the loft feature.

Creating the Extruded Feature

Follow these steps:

1. Right-click and select **New Sketch** from the marking menu.

2. Select the plane located at the top of the sweep feature.

3. Right-click and select **Center Point Circle** from the marking menu.

4. Select the origin point, move the cursor outside, and click to create a circle.

5. Right-click and select **OK.**

6. Right-click and select **General Dimension** from the marking menu.

7. Select the circle, move the cursor outward, and click.

8. Type **1.574** and press Enter.

9. Click **Finish Sketch** on the ribbon.

10. Click the **Extrude** button on the **Create** panel.

11. Extrude the circle up to 1 in.

Creating the Emboss Feature

Follow these steps:

1. Click **3D Model ➤ Work Features ➤ Plane ➤ Offset from Plane** on the ribbon.

2. Select the YZ plane from the Browser window.

3. Enter **2** in the **Distance** box and click **OK** .

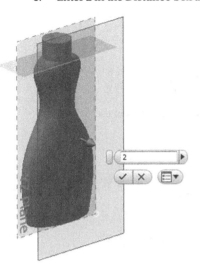

4. Click **3D Model ➤ Sketch ➤ Start 2D Sketch** on the ribbon.

5. Select the newly created datum plane.

6. Click **Sketch ➤ Create ➤ Circle ➤ Ellipse** on the ribbon.

7. Draw the ellipse by selecting the points, as shown here:

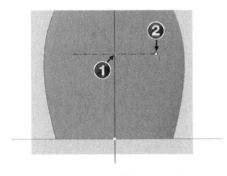

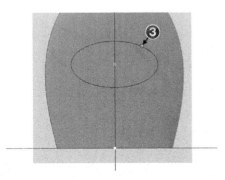

8. On the ribbon, click **Sketch ➤ Constrain ➤ Dimension**.

9. Select the ellipse, move the cursor downward, and click.

10. Type **2** and press Enter.

11. Select the ellipse, move the cursor toward the left, and click.

12. Type **1.35** and press Enter.

13. Select the origin point of the sketch and the center point of the ellipse.

14. Move the cursor toward the left and click to place the dimension between the selected points.

15. Type **2.55** and press Enter.

16. Right-click and select **Line** from the marking menu.

17. Right-click and select **Construction** from the marking menu.

18. Select the origin point of the sketch and the center point of the ellipse.

19. Right-click and select **OK**; a construction line is created between the sketch origin and the ellipse. Also, the sketch is fully constrained.

20. Click **Finish Sketch**.

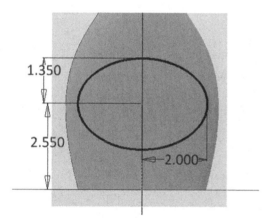

21. Click **3D Model ➤ Create ➤ Emboss** on the ribbon; the **Emboss** dialog appears.

22. Select the sketch, if not already selected.

23. Click the **Engrave from Face** button in the dialog.

24. Set **Depth** to 0.125.

25. Click **OK** to create the embossed feature.

Mirroring the Emboss Feature

Follow these steps:

1. Click **3D Model ➤ Pattern ➤ Mirror** on the ribbon.

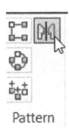

2. Select the emboss feature from the model geometry.

3. In the **Mirror** dialog, click the **Mirror Plane** button and then select the YZ plane from the Browser window.

4. Click **OK** to mirror the emboss feature.

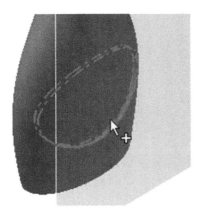

Creating Fillets

Follow these steps:

1. Click **3D Model ➤ Modify ➤ Fillet** on the ribbon; the **Fillet** dialog appears.

2. Click the bottom and top edges of the swept feature.

3. Set **Radius** to 0.2.

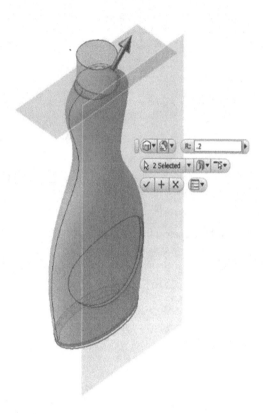

4. Click "**Click to add**" in the Fillet dialog.

5. Set **Radius** to 0.04.

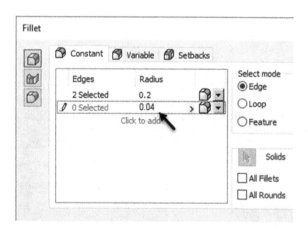

6.   Select the edges of the emboss features and click **OK**.

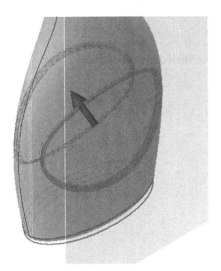

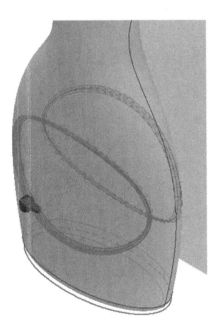

Shelling the Model

Follow these steps:

1. Click **3D Model ➤ Modify ➤ Shell** 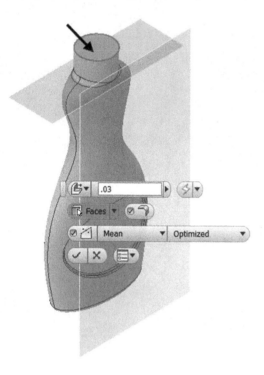 on the ribbon; the **Shell** dialog appears.

2. Set **Thickness** to 0.03.

3. Select the top face of the cylindrical feature.

4.   Click **OK** to create the shell.

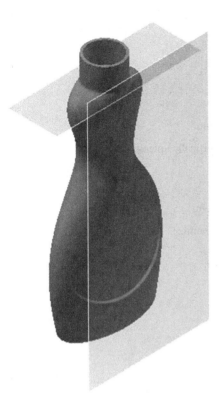

Adding Threads

Follow these steps:

1.   Select the YZ plane and then click **Create Sketch**.

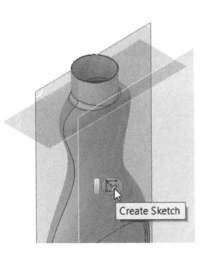

2. Click the **Slice Graphics** icon located at the bottom of the window.

3. On the ribbon, click **Sketch ➤ Create ➤ Line**.

4. Right-click and select **Centerline** from the marking menu.

5. Select the origin point of the sketch, move the cursor vertically upward, and click to create a vertical centerline. Press Esc.

6. Deactivate the **Centerline** icon on the **Format** panel.

7. Right-click and select **Create Line** from the marking menu.

8. Right-click and select **Construction** from the marking menu.

9. Create a horizontal construction line, as shown here:

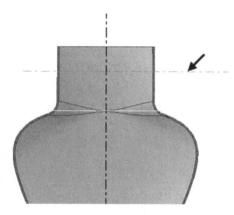

10. Deactivate the **Construction** icon on the **Format** panel of the **Sketch** ribbon tab.

11. On the ribbon, click **Sketch ➤ Create ➤ Line**.

12. Create a closed profile, as shown here:

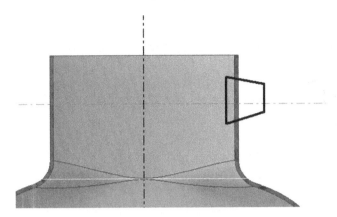

13. On the ribbon, click **Sketch ➤ Constrain ➤ Symmetric** ⊏⌐⊐ .

14. Select the two inclined lines of the sketch and then select the construction line; the two inclined lines are made symmetric about the construction line.

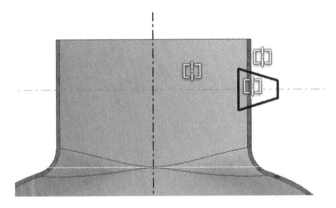

15. Click the **Dimension** button on the **Constrain** panel.

16. Select the two inclined lines, move the cursor horizontally toward the left, and click.

17. Type **60** in the **Edit Dimension** box and press Enter.

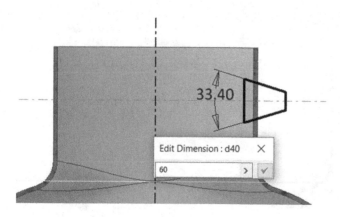

18. Draw the thread profile.

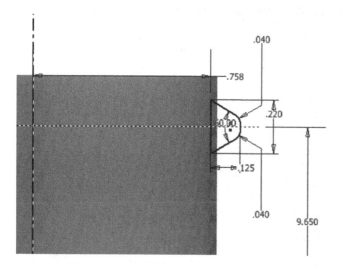

19. On the ribbon, click **Sketch ➤ Create ➤ Fillet** .

20. Type **0.04** in the **2D Fillet** box.

21. Make sure that the **Equal** button is active on the **2D Fillet** box.

22.   Select the two corners of the sketch, as shown here:

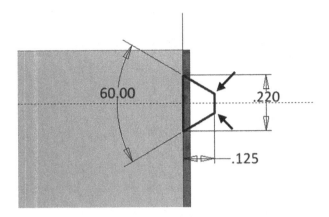

23.   Close the **2D Fillet** dialog box.

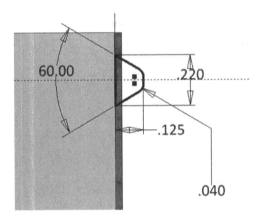

24.   Click **Finish Sketch** on the ribbon.

25.   On the ribbon, click **3D Model ➤ Create ➤ Coil**   ; the closed profile of the sketch is selected automatically.

26. Select the axis of the coil.

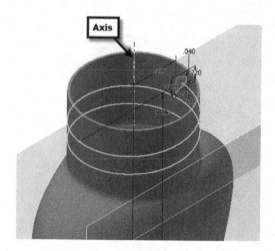

27. In the dialog, click the **Coil Size** tab and select **Type ➤ Pitch and Revolution**.

28. Type **0.275** and **2** in the **Pitch** and **Revolution** boxes, respectively.

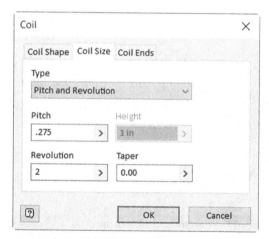

29.   Click **OK**.

30.   In the Browser window, right-click the YZ plane and then select **New Sketch**.

31.   On the ribbon, click **Sketch ➤ Create ➤ Project Cut Edges ➤ Project Geometry** .

32.   Select the edges of the end face of the thread.

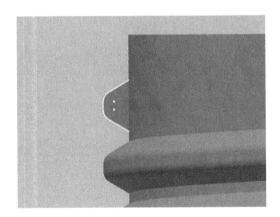

33. Draw a straight line connecting the end points of the projected elements.

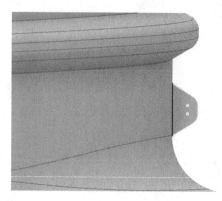

34. Click **Finish Sketch**.

35. Activate the **Revolve** tool and click the vertical line of the sketch.

36. In the dialog, select **Extents ➤ Angle** and then type **100** in the **Angle1** box.

37. Click the **Direction 1** button.

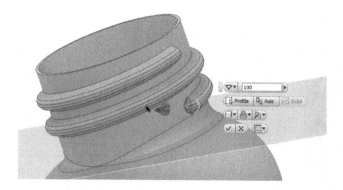

38. Click **OK**.

39. Likewise, blend the other end of the thread. Note that you need to reverse the direction of revolution

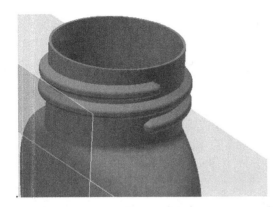

40. Save the model.

Tutorial 5

In this tutorial, you will create a chair, as shown here:

Creating a 3D Sketch

Follow these steps:

1. Open a new Inventor file using the Standard.ipt template (see Chapter 2's Tutorial 3).

2. Click the **Home** icon located above the ViewCube. This changes the view orientation to Home.

3. On the ribbon, click **3D Model ➤ Sketch ➤ Start 2D Sketch ➤ Start 3D Sketch**.

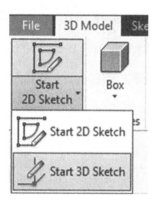

4. On the **3D Sketch** tab of the ribbon, click **Draw ➤ Line**.

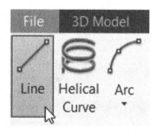

5. Expand the **Draw** panel on the ribbon and activate the **Precise Input** option.

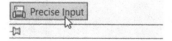

6. On the **Precise Input** toolbar, click the **Reset to Origin** button.

7. Select the **Relative** option from the drop-down available on the **Precise Input** toolbar.

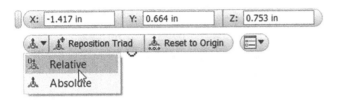

8. On the **Precise Input** toolbar, click in the X box and type **0**.

9. Press the Tab key and type **0** in the Y box.

10. Press the Tab key and type **0** in the Z box.

11. Press Enter to specify the first point.

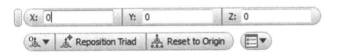

12. Type **12** in the **X** box and press Tab on your keyboard.

13. Likewise, type **0** in the **Y** and **Z** boxes. Press Enter to specify the second point.

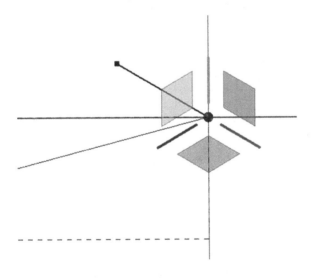

14. Type **0**, **0**, and **20** in the X, Y, and Z boxes, respectively. Press Enter.

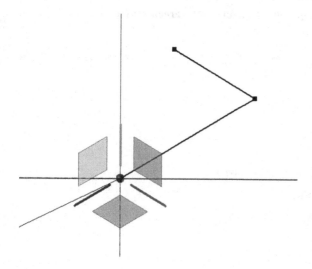

15. Type **0**, **18**, **0** in the X, Y, and Z boxes, respectively. Press Enter.

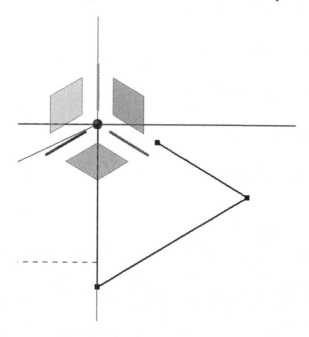

16. Type **0**, **0**, and **-22** in the X, Y, and Z boxes, respectively. Press Enter.

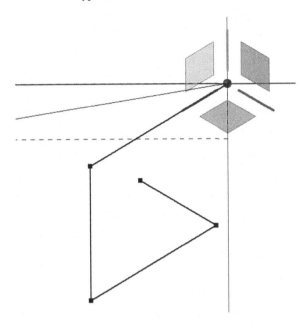

17. Type **0**, **18**, and **0** in the X, Y, and Z boxes, respectively. Press Enter.

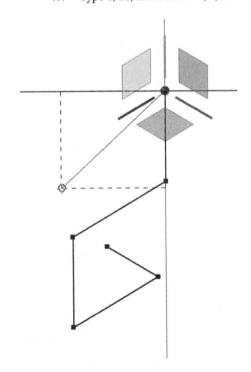

18. Type **-12**, **0**, and **0** in the X, Y, and Z boxes, respectively. Press Enter.

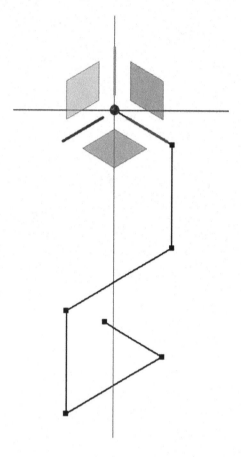

19. Right-click and select **OK**.

20. On the **3D Sketch** tab of the ribbon, click **Pattern ➤ Mirror**.

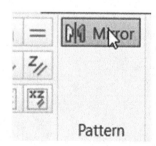

21.   Drag a selection box and select all the sketch elements.

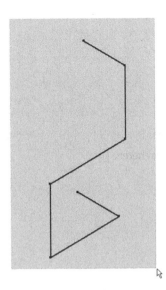

22.   Click the **Mirror Plane** button in the dialog and then select YZ Plane from the Browser window.

23.   Click **Apply** and **Done** in the dialog.

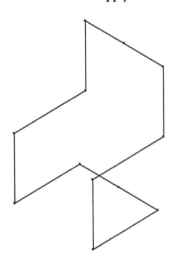

24. On the ribbon, click **3D Sketch ➤ Draw ➤ Bend**.

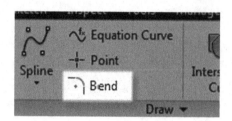

25. Type **3** in the **Bend** dialog and select the intersecting lines, as shown here:

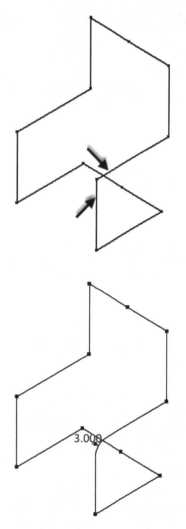

26.   Likewise, bend the other corners of the 3D sketch.

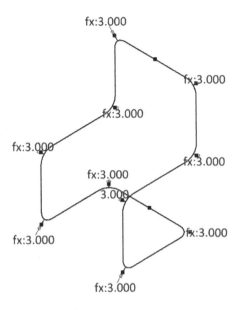

27.   Right-click and select **OK**.

28.   On the ribbon, click **3D Sketch ➤ Constrain ➤ Fix** 🔒 .

29.   Select the origin point of the sketch.

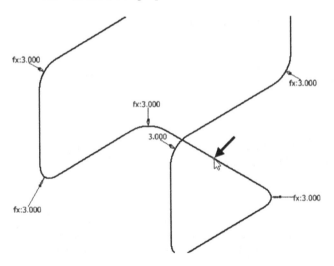

30. Add dimensions to fully define the sketch.

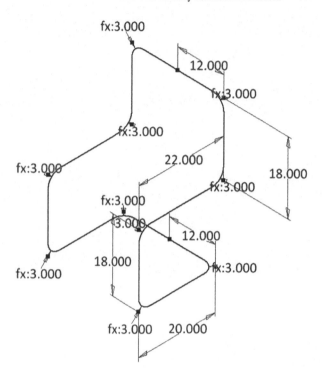

31. Click **Finish Sketch** on the ribbon.

32. On ribbon, click **3D Model ➤ Work Features ➤ Plane ➤ Normal to Axis through Point**.

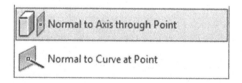

33. Click the horizontal line of the sketch and its endpoint.

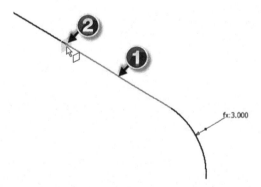

34.   Click **3D Model ➤ Sketch ➤ Start 2D Sketch** on the ribbon.

35.   Select the newly created plane.

36.   Create two concentric circles of 1.250 and 1 diameters.

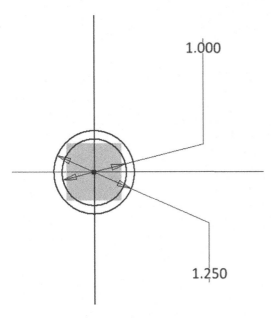

37.   Add dimensions from the sketch origin to position the circles.

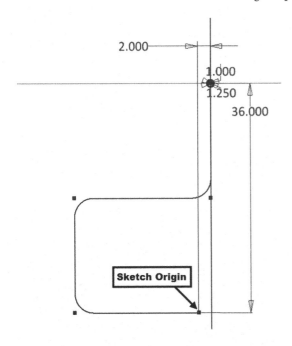

38.   Click **Finish Sketch**.

Creating the Sweep Feature

Follow these steps:

1. On the ribbon, click **3D Model ➤ Create ➤ Sweep**.

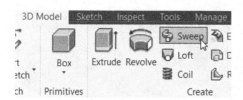

2. Zoom into the circular sketch and click in the outer loop.

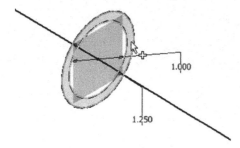

3. Click the 3D sketch to define the sweep path.

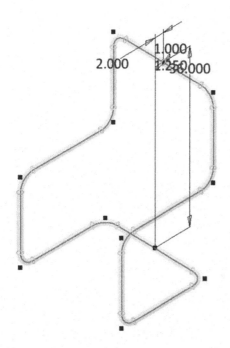

4.   Click **OK** to sweep the profile.

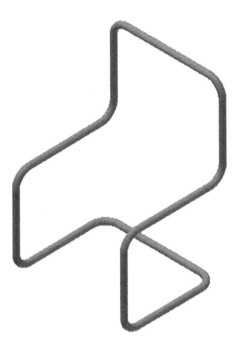

5.   Start a sketch on the YZ plane.

6.   Click the **Slice Graphics** 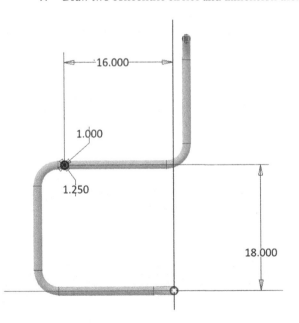 icon at the bottom of the graphics window.

7.   Draw two concentric circles and dimension them.

8. Click **Finish Sketch**.

9. Activate the **Extrude** tool and click in the outer loop of the sketch.

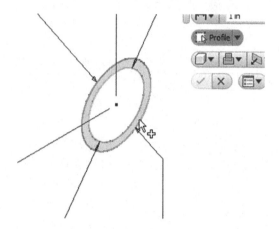

10. In the **Extrude** dialog, select **Extents ➤ Between**.

11. Click the tubes on both sides of the sketch.

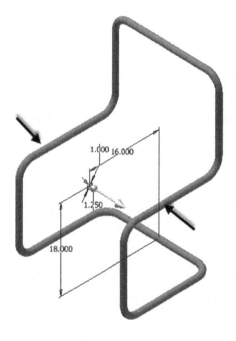

12. In the **Extrude** dialog, deselect the **Check to Terminate feature in the extended face** options.

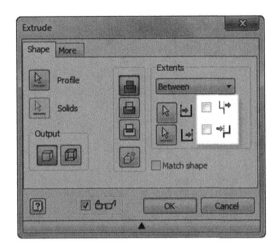

This will avoid the extruded feature from terminating on the extended portion of the selected surface.

13. Click **OK** to extrude the sketch.

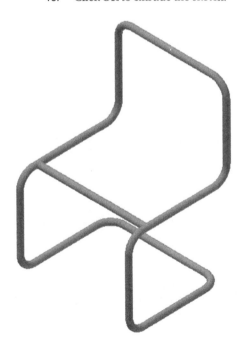

Creating the Along Curve Pattern

Follow these steps:

1. In the Browser window, right-click **3D Sketch** and select **Visibility**; the 3D sketch is displayed.

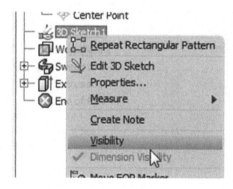

2. On the ribbon, click **3D Model ➤ Pattern ➤ Rectangular Pattern**.

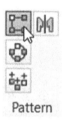

3. Click the extrude feature.

4. In the **Rectangular Pattern** dialog, click the **Direction 1** button and then click the **3D sketch**.

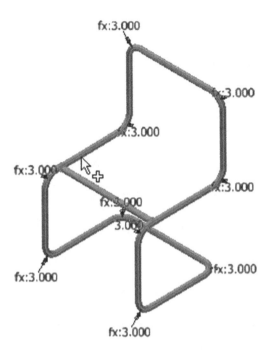

5. Type **3** and **23** in the **Column Count** and **Column Spacing** boxes, respectively.

6. Select **Spacing ➤ Distance** from the drop-down menu.

7. Click the double-arrow button located at the bottom of the dialog. This expands the dialog.

8.   Set **Orientation** to **Direction 1.**

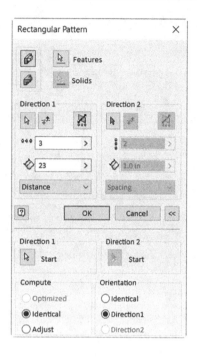

9.   Click **OK** to pattern the extruded feature.

10. In the Browser window, right-click 3D Sketch and select **Visibility**. This hides the 3D sketch.

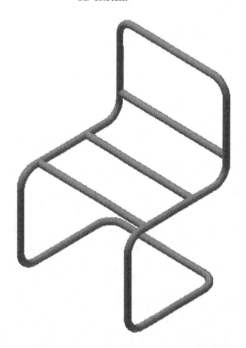

Creating the Free-Form Feature

Follow these steps:

1. On the ribbon, click **3D Model ➤ Work Features ➤ Plane**.

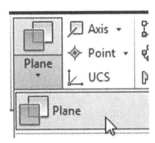

2. In the **Browser window**, click XZ Plane. A plane appears on the XZ plane.

3. Click the top portion of the extruded feature. A plane appears tangent to the extruded feature.

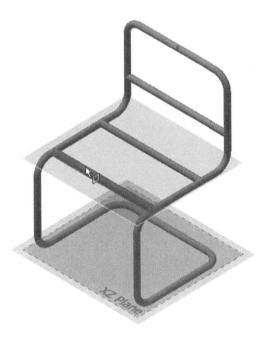

4. Start a sketch on the new plane.

5. Place a point on the sketch plane and add dimensions to position it.

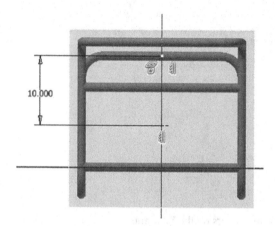

6. Click **Finish Sketch**.

7. On the ribbon, click **3D Model ➤ Freeform ➤ Box** .

8. Select the plane tangent to the extruded feature.

9. Select the sketch point to define the location of the free-form box.

10. Click and drag the side arrow of the free-form box.

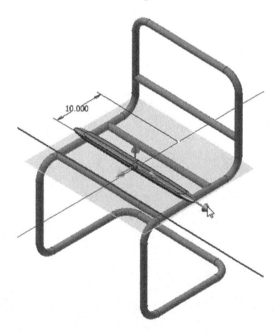

11.   Click and drag the front arrow of the free-form box.

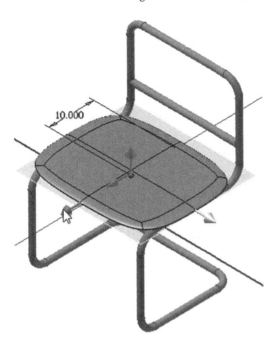

12.   Click and drag the top arrow to increase the height of the free-form box.

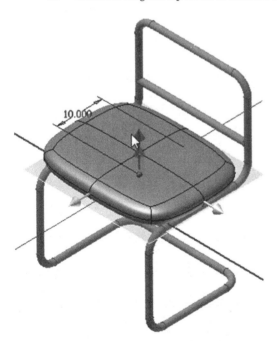

13. In the **Box** dialog, type **4**, **1**, and **2** in the **Faces** boxes, respectively.

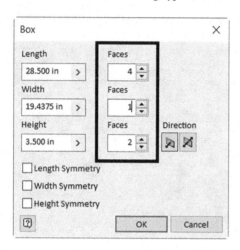

14. Click **OK** to create the free-form shape.

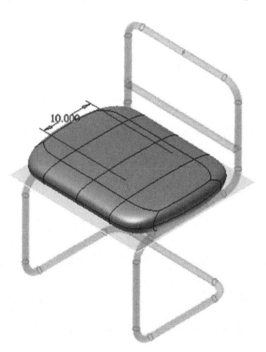

Editing the Free-Form Shape

Follow these steps:

1. On the ribbon, click **Freeform ➤ Edit ➤ Edit Form** .

2. Hold the Ctrl key and click the top faces of the free-form shape.

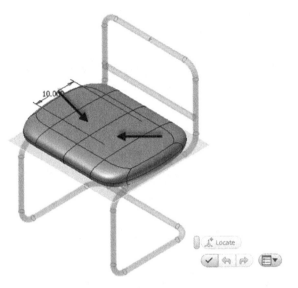

3. Click the arrow-pointing upward.

4. Drag it downward 1.6875 in, as shown here:

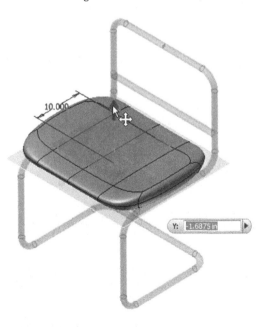

5.   Click **OK** in the dialog.

6.   On the ribbon, click **Freeform ➤ Edit ➤ Edit Form** .

7.   Hold the Ctrl key and click the two edges at the front.

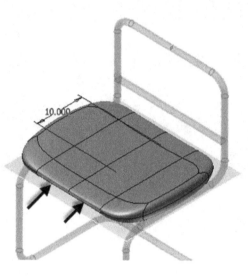

8.   Drag the vertical arrow downward.

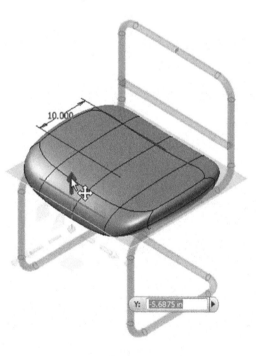

9. Click **OK** in the **Edit Form** dialog.

10. Click **Finish Freeform** on the ribbon.

Create Another Free-Form Box

Follow these steps:

1. On the ribbon, click **3D Model** ➤ **Work Features** ➤ **Plane**  .

2. In the Browser window, click XY Plane.

3. Click the vertical portion of the sweep feature to create a plane tangent to it.

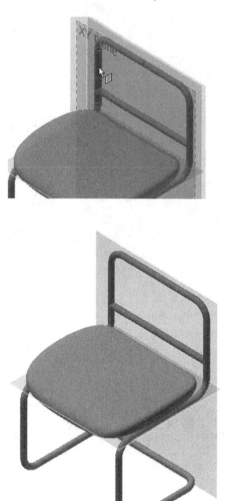

4. Start a sketch on the new plane.

5. Place a point and add dimensions to it.

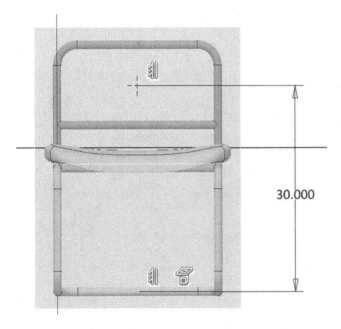

30.000

6. Click **Finish Sketch**.

7. Activate the Freeform **Box** 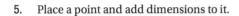 tool.

8. Select the new plane and click the sketch point.

9. In the **Box** dialog, type **27**, **16**, and **3** in the **Length, Width,** and **Height** boxes, respectively.

10.   Click **OK** and then click **Finish Freeform**.

11.   Save and close the file.

Tutorial 6

In this tutorial, you will create a bolt.

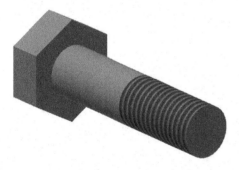

Starting a New Part File

Follow these steps:

1. Start a new part file using the **Standard.ipt** template (see Chapter 2's Tutorial 3).

2. On the ribbon, click **3D Model ➤ Primitives ➤ Primitive drop-down ➤ Cylinder** .

3. Click the YZ plane.

4. Click the origin point of the sketch to define the center point of the circle.

5. Move the pointer and type **0.75** in the box and then press Enter.

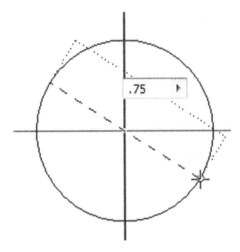

6. Type **3** in the **Distance** box and press Enter.

219

Creating the Second Feature

Follow these steps:

1. Start a sketch on the YZ plane.

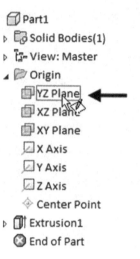

2. On the ribbon, click **Sketch ➤ Create**, click the **Rectangle** drop-down, and select **Polygon**.

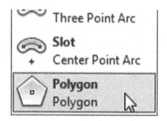

3. Click the sketch origin.

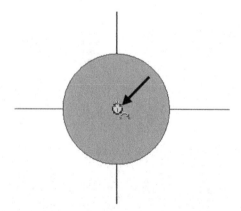

4. Type **6** in the **Polygon** dialog.

5. Move the pointer vertically upward. You will notice that a dotted trace line appears between the origin point and the pointer.

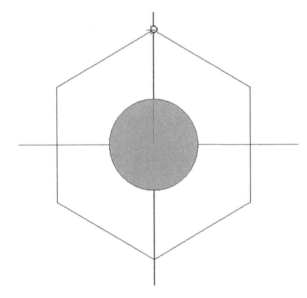

6. Click to create the polygon.

7. Click **Done** in the dialog.

8. On the ribbon, click **Sketch ➤ Format ➤ Construction** .

9. Activate the **Line** tool and select the vertices of the polygon.

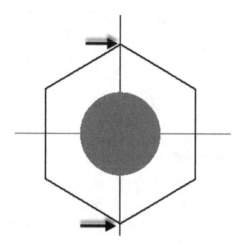

10. Activate the **Dimension** tool and create a dimension, as shown here:

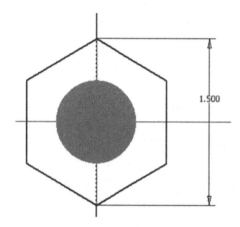

11. Finish the sketch.

12. Activate the **Extrude** tool and select the sketch, if not already selected.

13. In the **Extrude** dialog, type **0.5** in the Distance box.

14. Use the **Direction 1** or **Direction 2** button to make sure that the polygon is extruded toward the left.

15.   Click **OK**.

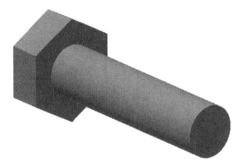

Adding Threads

Follow these steps:

1.   On the ribbon, click **3D Model ➤ Modify ➤ Thread**  .

2.   Click the cylindrical face of the model geometry.

3.   In the **Thread** dialog, uncheck the **Full Length** option and type **1.5** in the
      Length box.

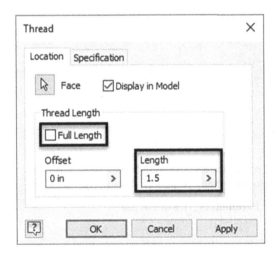

4.   Click the **Specification** tab to specify the thread settings.

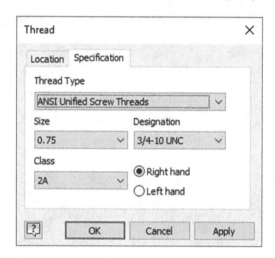

5.   Click **OK** to add the thread.

Creating iParts

The iParts feature allows you to design a part with different variations, sizes, materials, and other attributes. Now, you will create different variations of the bolt created in the previous section.

1.   On the ribbon, click **Manage ➤ Parameters ➤ Parameters**. This opens the **Parameters** dialog.

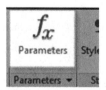

2. In the **Parameters** dialog, click in the first cell of the **Model Parameters** table and type **Diameter**.

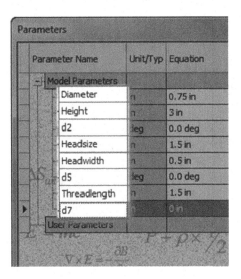

3. Likewise, change the names of other parameters.

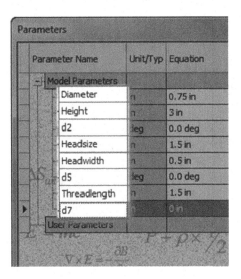

4. Click **Done** in the **Parameters** dialog.

5. Right-click in the graphics window and select **Dimension Display ➤ Expression**.

6. In the Browser window, right-click **Extrusion1** and select **Show Dimensions**. You will notice that the dimensions are shown along with the names.

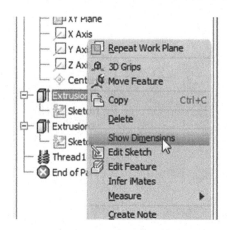

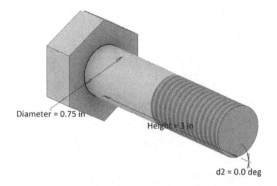

7. On the ribbon, click **Manage ➤ Author ➤ Create iPart**.

This opens the **iPart Author** dialog. In this dialog, you will define the parameters to create other versions of the model geometry.

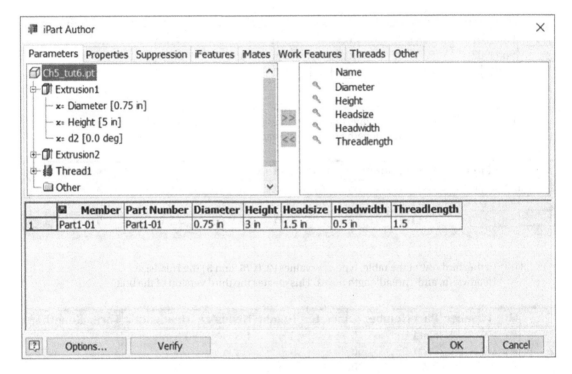

The table at the bottom of this dialog shows the parameters for the iPart factory. You will notice that the renamed parameters are automatically added to the table. If you want to add more parameters to the table, then select them from the section located at the left side. Click the arrow button pointing toward the right. Likewise, if you want to remove a parameter from the table, then select it from the right-side section and click the arrow pointing toward the left.

8. Now, right-click the table and select **Insert Row**. Notice that a new row is added to the table.

	💾 Member	Part Number	Diameter	Height
1	Part5-01	Part5-01	0.75 in	3 in
		Insert Row ⬅		
		Delete Row		
		Set As Default Row		

9. Likewise, insert another row.

	💾 Member	Part Number	Diameter	Height	Headsize	Headwidth	Threadlength
1	Part1-01	Part1-01	0.75 in	3 in	1.5 in	0.5 in	1.5
2	Part1-02	Part1-02	0.75 in	3 in	1.5 in	0.5 in	1.5
3	Part1-03	Part1-03	0.75 in	3 in	1.5 in	0.5 in	1.5

10. In the second row of the table, type new values (**5**, **0.75**, and **3**) in the Height, Headwidth, and Threadlength boxes. This creates the second version of the bolt.

	💾 Member	Part Number	Diameter	Height	Headsize	Headwidth	Threadlength
1	Part1-01	Part1-01	0.75 in	3 in	1.5 in	0.5 in	1.5
2	Part1-02	Part1-02	0.75 in	5	1.5 in	.75	3
3	Part1-03	Part1-03	0.75 in	3 in	1.5 in	0.5 in	1.5

11. In the third row of the table, type new values (**2**, **0.75**, and **3**) the Headsize, Headwidth, and Threadlength boxes. This creates the third version of the bolt.

	💾 Member	Part Number	Diameter	Height	Headsize	Headwidth	Threadlength
1	Part1-01	Part1-01	0.75 in	3 in	1.5 in	0.5 in	1.5
2	Part1-02	Part1-02	0.75 in	5	1.5 in	.75	3
3	Part1-03	Part1-03	0.75 in	3 in	2	.75	3

Now, you have to set the default version of the bolt.

12. Right-click the third row of the table and select **Set As Default Row**.

2	Part5-02	Part5-02	0.75 in	5	1.5 in
3	Part5-03	Part5-03	0.75 in	3 in	2

Insert Row

Delete Row

Set As Default Row ⬅

Options..

13. Click **OK** to close the dialog. Notice that the default version of the bolt changes.

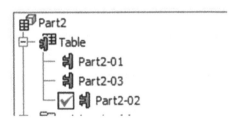

In the Browser Bar, you will notice that the **Table** item is added.

14. Expand the **Table** item in the Model Window to view the different variations of the iPart. Notice that the activated version of the iPart is designated by a check mark.

Part2
└ Table
   ├ Part2-01
   ├ Part2-03
   └ ✓ Part2-02

15.   Double-click any other version of the iPart to activate it.

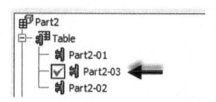

If you want to make changes to any version of the bolt, then right-click it and select **Edit Table**.

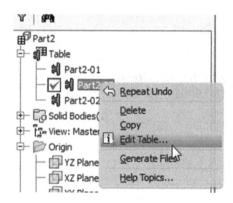

If you want to edit the table using a spreadsheet, then right-click **Table** and select **Edit via Spreadsheet**. Click **OK** on the message box.

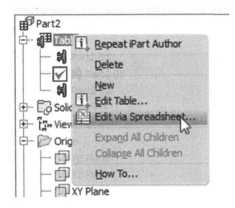

Now, modify the values in the spreadsheet and close it. A message pops up asking you to save the changes. Click **Save** to save the changes.

	A	B	C	D	E	F	G	H
1	Member<	Part Numl	Diameter	Height	Headsize	Headwidt	Threadler	Thread1
2	Part2-01	Part2-01	0.75 in	3 in	1.5 in	0.5 in	1.5 in	Compute
3	Part2-03	Part2-03	0.75 in	5	1.5 in	0.75 in	3	Compute
4	Part2-02	Part2-02	0.75 in	3 in	2	0.75 in	3	Compute

If you want to save any iPart version as a separate part file, then right-click it and select **Generate Files**.

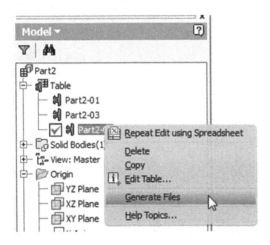

16.   Save and close the file.

Tutorial 7

In this tutorial, you will create a plastic casing.

Creating the First Feature

Follow these steps:

1. Open a new Autodesk Inventor part file using the **Standard.ipt** template (see Chapter 2's Tutorial 3).

2. On the ribbon, click **3D Model ➤ Sketch ➤ Start 2D Sketch**.

3. Select the XZ plane.

4. On the ribbon, click **Sketch ➤ Create ➤ Line**.

5. Click in the second quadrant, move the cursor horizontally toward the right, and then click to create a horizontal line, as shown here:

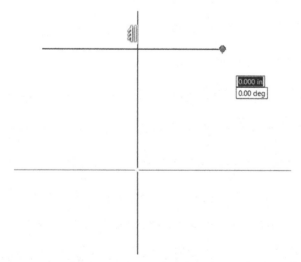

6. On the ribbon, click **Sketch ➤ Create**, click the **Arc drop-down**, and select **Arc Three Point**.

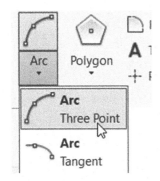

7. Select the right endpoint of the horizontal line.

8. Move the cursor vertically downward and click to define the second point.

9. Move the pointer toward the right and click the horizontal axis line.

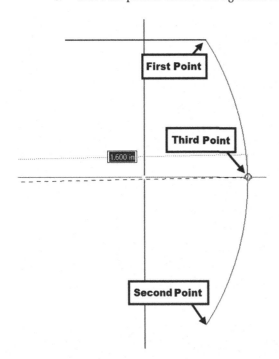

10. Likewise, create another three-point arc and horizontal line, as shown here:

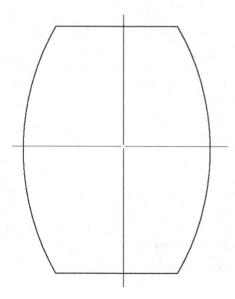

11. Create a vertical construction from the origin point.

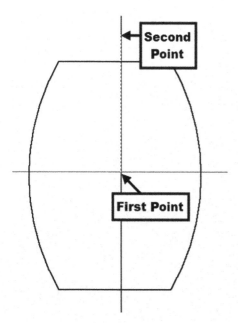

12. On the ribbon, click **Sketch ➤ Constrain ➤ Symmetric** 〖◻〗.

13. Select the two arcs and the construction.

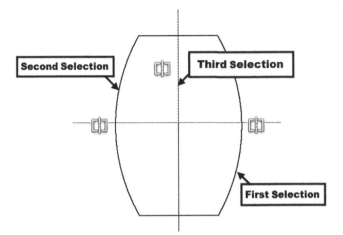

14. Likewise, create a horizontal construction line from the origin point and then make the two horizontal lines symmetric about it.

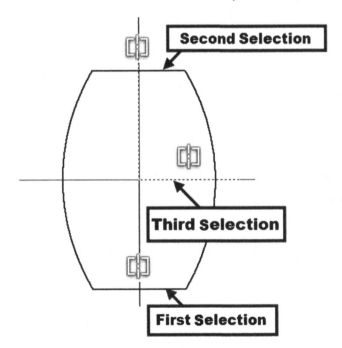

15. Add dimensions to the sketch. Also, add the vertical constraint between the origin and center point of the arc.

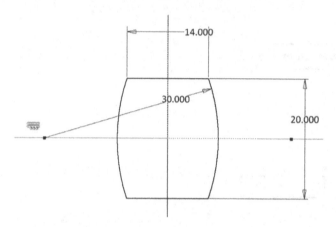

16. Click **Finish Sketch**.

17. Click **3D Model ➤ Create ➤ Extrude** on the ribbon; the **Extrude** dialog appears.

18. Set **Distance** to 3.15.

19. Click the **More** tab and set the **Taper** angle to -10.

20. Click the **OK** button.

Creating the Extruded Surface

Follow these steps:

1. Click **3D Model ➤ Sketch ➤ Start 2D Sketch** on the ribbon and then select the XY plane.

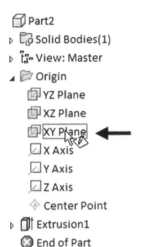

2. Click the **Slice Graphics** button at the bottom of the window or press **F7** on the keyboard.

3. Click **Sketch ➤ Create ➤ Line ➤ Spline Interpolation** on the ribbon.

4. Create a spline, as shown here (see Chapter 5's Tutorial 4):

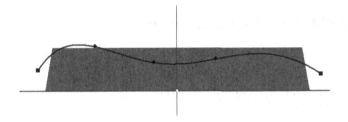

5. Apply dimensions to the spline, as shown here:

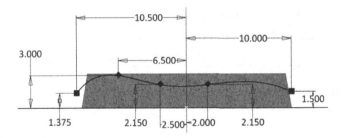

6. Click **Finish Sketch**.

7. Click the **Extrude** button.

8. Select the sketch, if not already selected.

9. In the **Extrude** dialog, set the **Output** type to **Surface**.

10. Set the **Extents** type to **Distance**.

11. Select the **Symmetric** button.

12. Extrude the sketch up to a distance of 17 in.

Replacing the Top Face of the Model with the Surface

Follow these steps:

1. On the **Surface** panel of the **3D Model ribbon**, click the **Replace Face** button; the **Replace Face** dialog appears.

   Now, you need to select the face to be replaced.

2. Select the top face of the model.

   Next, you need to select the replacement face or surface.

3. Click the **New Faces** button in the dialog and select the extruded surface.

 You can also use a solid face to replace an existing face.

---

4. Click **OK** to replace the top face with a surface.

5. Hide the extruded surface by right-clicking it in the Browser window and deselecting **Visibility**.

Creating a Face Fillet

Follow these steps:

1. Click the **Fillet** button on the **Modify** panel.

2. Click the **Face Fillet** button in the **Fillet** dialog.

3. Select the top surface as the first face and the inclined front face as the second face.

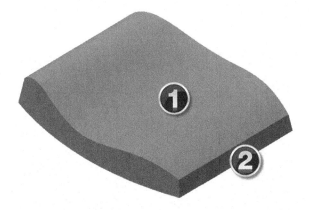

4.   Set **Radius** to 1.5 and click the **OK** button to create the face fillet.

5.   Likewise, apply a face fillet of 1.5 radius between the top surface and the back inclined face of the model.

Creating a Variable Radius Fillet

Follow these steps:

1.   Click the **Fillet**  button on the **Modify** panel.

2. Click the **Variable** tab in the **Fillet** dialog.

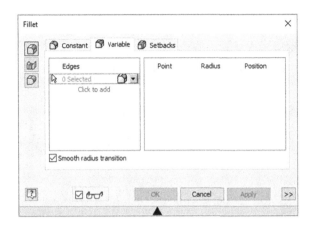

3. Select the curved edge on the model; the preview of the fillet appears.

4. Select a point on the fillet, as shown here:

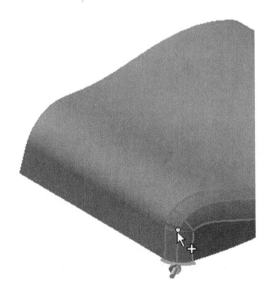

5. Select another point on the fillet, as shown here:

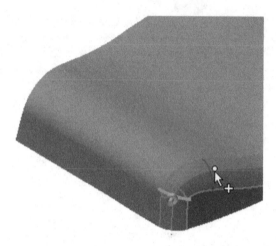

6. Set the radii of the **Start, End, Point 1,** and **Point 2**, as shown here:

Point	Radius	Position
Start	.6	0.0
End	.6	1.0
Point 1	1	0.0000
Point 2	.8	0.9103

You can also specify the fillet continuity type. By default, the **Tangent Fillet** type is specified.

7. Select **Smooth (G2) Fillet** type from the **Edges** section.

8. Make sure that the **Smooth radius transition** option is checked.

9. Click **OK** to create the variable fillet.

Mirroring the Fillet

Follow these steps:

1. Click the **Mirror**  button on the **Pattern** panel; the **Mirror** dialog appears.

2. Select the variable radius fillet from the model.

3. Click the **Mirror Plane** button in the dialog.

4. Select **XY Plane** from the Browser window.

5. Click **OK** to mirror the fillet.

Shelling the Model

Follow these steps:

1. Click the **Shell** button on the **Modify** panel; the **Shell** dialog appears.

2. Click the **Inside** button in the dialog and set **Thickness** to 0.2 in.

3. Rotate the model and select the bottom face.

4. Click **OK**.

Creating the Boss Features

Follow these steps:

1. Click **3D Model ➤ Sketch ➤ Start 2D Sketch** on the ribbon and select the bottom face of the model.

2. Activate the **Construction** button on the **Format** panel.

3. On the ribbon, click **Sketch ➤ Create**, select the **Rectangle drop-down**, and select **Rectangle Two Point Center**.

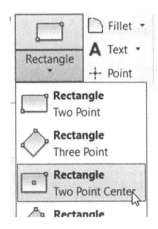

4. Select the origin point of the sketch.

5. Move the cursor outward and click to create the rectangle.

6. Apply dimensions to the rectangle.

7. Click the **Point** button on the **Create** panel.

8. Place four points at the corners of the rectangle.

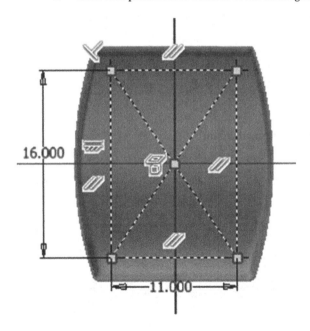

9. Click **Finish Sketch**.

Now, you will create bosses by selecting the points created in the sketch.

10. On the ribbon, click the **Show Panels** ⊙ ▾ button located on the right side and then select **Plastic Part** from the menu.

11. Click the **Boss** button on the **Plastic Part** panel; the **Boss** dialog appears.

12. Click the **Thread** button in the dialog.

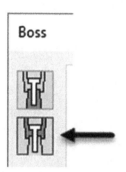

13. Select the **From Sketch** option from the **Placement** group.

14. Select the points located on the corners of the rectangle, if not already selected; the bosses are placed at the selected points.

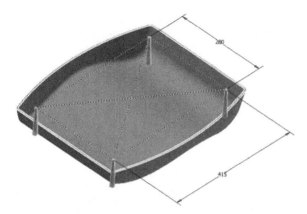

15. Click the **Thread** tab and specify the parameters, as shown here:

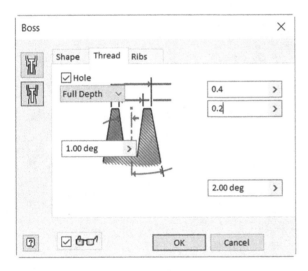

16. Click the **Ribs** tab and check the **Stiffening Ribs** option.

17. Set the rib parameters, as shown here:

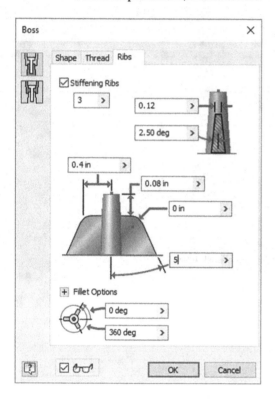

18. Expand the **Fillet options**.

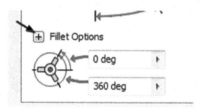

19. Specify the fillet options, as shown here:

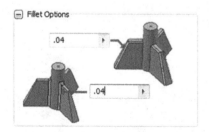

20.   Click **OK** to create the bosses with ribs.

Creating the Lip Feature

Follow these steps:

1.   Click the **Lip** 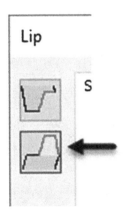 button on the **Plastic Part** panel of the ribbon; the **Lip** dialog appears.

2.   Click the **Lip** button in the dialog.

3. Select the outer edge of the bottom face.

4. Click the **Guide Face** button in the dialog and select the bottom face of the model.

5.   Click the **Lip** tab and set the parameters, as shown here:

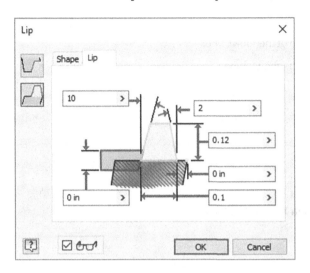

6.   Click **OK** to create the lip.

Creating the Grill Feature

Follow these steps:

1. Click the **Home** button located at the top of the ViewCube.

2. Click the corner point of the ViewCube, as shown here:

3. On the ribbon, click **3D Model ➤ Sketch ➤ Start 2D Sketch**.

4. Select the inclined face, as shown here:

5. Create the sketch using the **Rectangle Two Point Center** and **Line** tools.

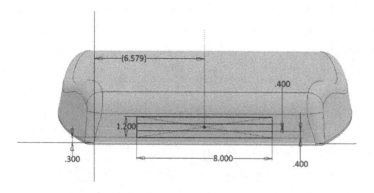

6. Click **Finish Sketch**.

7. Click the **Grill** button on the **Plastic Part** panel.

8. Select the rectangle as the boundary and set the **Boundary** parameters, as shown here:

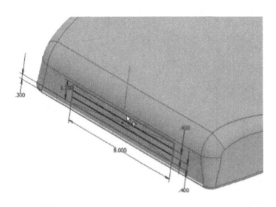

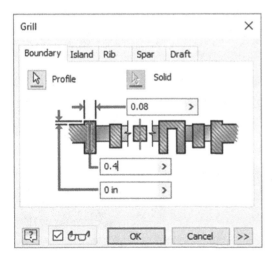

9. Click the **Rib** tab and select the horizontal lines.

10. Set the rib parameters, as shown here:

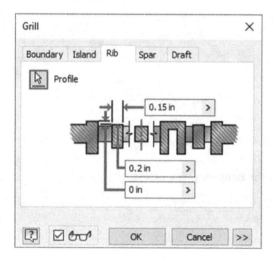

11. Click **OK** to create the grill

12. Save the model as Plastic Cover.ipt.

Creating Ruled Surface

Follow these steps:

1. Click **3D Model ➤ Surface ➤ Ruled Surface** 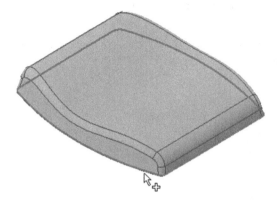 on the ribbon and select the bottom edge of the model.

2. Click the **Normal** button in the **Ruled Surface** dialog.

   The preview of the ruled surface appears normal to the selected edge.

   You can click the **Alternate All Faces** button to change the direction of the ruled surface.

3. Type **2** in the **Distance** box.

4. Click **OK** to create the ruled surface.

   The ruled surface can be used as a parting split while creating a mold.

5. Close the part file without saving.

Tutorial 8

In this tutorial, you will learn to extrude a sketch from a face.

1. Download Tutorial_8.ipt from the companion web site: www.apress.com.

2. Open the downloaded file.

3. On the ribbon, click the **3D Model** tab, select the **Create panel**, and click

    Extrude .

4. Click in the region of the sketch, as shown here:

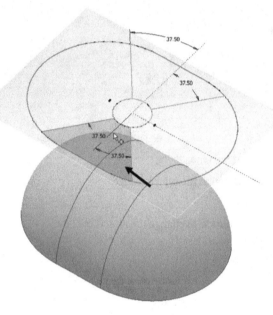

5. In the **Extrude** dialog, type **5** in the **Distance1** box located in the Extents section.

6. In the **Extrude** dialog, select the **Extents** drop-down and select **Distance from Face**.

7. Select the curved face, as shown here:

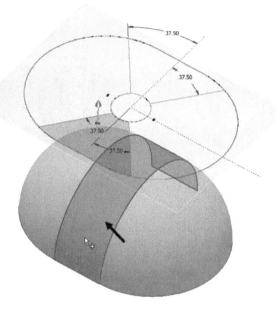

The sketch region is extruded from the selected face.

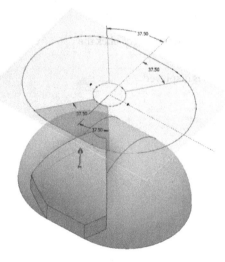

8. In the **Extrude** dialog, uncheck the **Select to Terminate feature by extending the face** ⊔⃗ option.

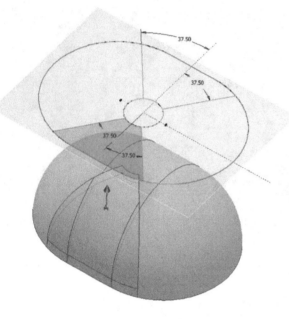

9. Click **OK** in the **Extrude** dialog.

10. In the Browser window, expand the Extrusion feature and then right-click the sketch.

11. Select **Visibility** from the shortcut menu; the sketch is displayed.

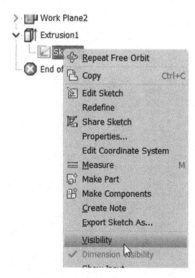

12. Likewise, extrude the other two sketch regions, as shown here. Use the **Distance from Face** option to define the extent.

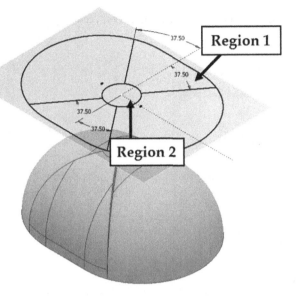

The Extrusion distances are 5 and 10 for region 1 and region 2, respectively.

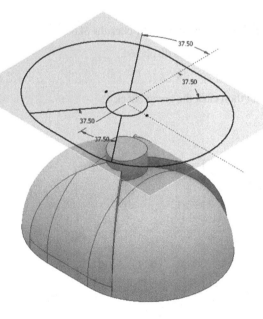

13. Save and close the part file.

Tutorial 9

In this tutorial, you will learn how to use the Extent Start option in the **Hole** tool.

1. Download Tutorial_9.ipt from the companion web site: www.apress.com.

2. Open the downloaded file.

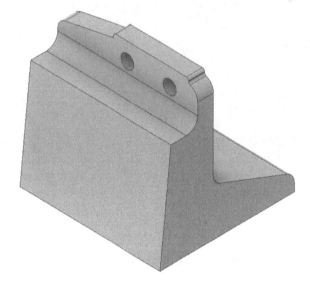

3. In the Browser window, expand the Hole feature, and then right-click the Sketch.

4. Select **Edit Sketch** from the shortcut menu; the sketch is displayed.

5. Double-click the 0.19 dimension.

6. Type **0.12** in the **Edit Dimension** box and then click the green check.

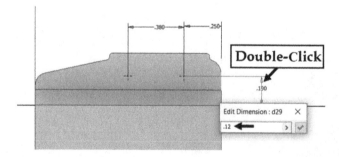

7. Click **Finish Sketch** on the ribbon.

    Notice that the fillet overlaps with the holes.

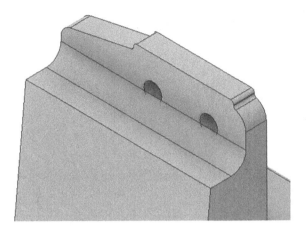

8. In the Browser window, right-click the **Hole** feature and then select **Edit Feature**.

9. In the **Hole** dialog, select the **Extent Start** option located at the bottom-right corner.

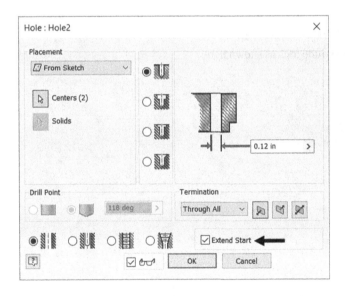

10. Click **OK** to close the dialog.

The Extent Start option removes the overlapping material.

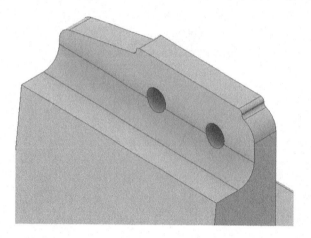

Tutorial 10

In this tutorial, you will learn to create partial chamfer.

1. Start a new part file using the **Standard(in).ipt** template.

2. Create a 1 × 1 × 1 box using the Extrude tool, as shown here:

3.	On the ribbon, click the **3D Model tab**, select the **Modify** panel, and click **Chamfer**.

4.	In the **Chamfer** dialog, click the **Partial** tab and then click the selected edge, as shown here:

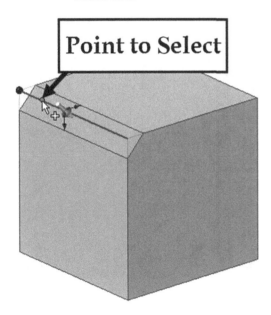

5.	On the **Partial** tab, select **Set the Driven Dimension ➤ To End**.

6.	Change the **To Start** and **Chamfer** values to 0.25 and 0.5, respectively.

7.   Click **OK**.

8.   Save and close the part file.

CHAPTER 6

■ ■ ■

Sheet Metal Modeling

This chapter will show you how to do the following:

- Create a face feature
- Create a flange
- Create a contour flange
- Create a corner seam
- Create punches
- Create a bend feature
- Create corner rounds
- Create a flat pattern

Tutorial 1

In this tutorial, you will create the sheet metal model shown here:

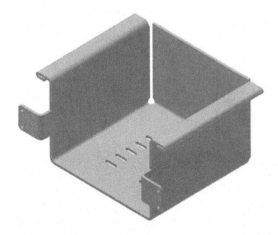

Starting a New Sheet Metal File

Follow these steps:

1. To start a new sheet metal file, click **Get Started Launch ➤ New** on the ribbon.

2. In the **Create New File** dialog, click the **Sheet Metal.ipt** icon and then click **Create**.

Sheet Metal.ipt

Setting the Parameters of the Sheet Metal Part

Follow these steps:

1. To set the parameters, click **Sheet Metal ➤ Setup ➤ Sheet Metal Defaults** on the ribbon; the **Sheet Metal Defaults** dialog appears.

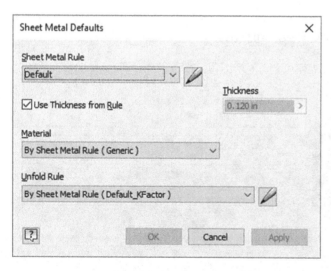

This dialog displays the default preferences of the sheet metal part such as sheet metal rule, thickness, material, and unfold rule. You can change these preferences as per your requirements.

2.   To edit the sheet metal rule, click the **Edit Sheet Metal Rule** button in the dialog.

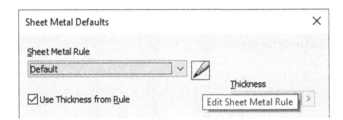

In the **Sheet** tab of the **Style and Standard Editor** dialog, you can set the
sheet preferences such as sheet thickness, material, flat pattern bend angle
representation, flat pattern punch representation, and gap size.

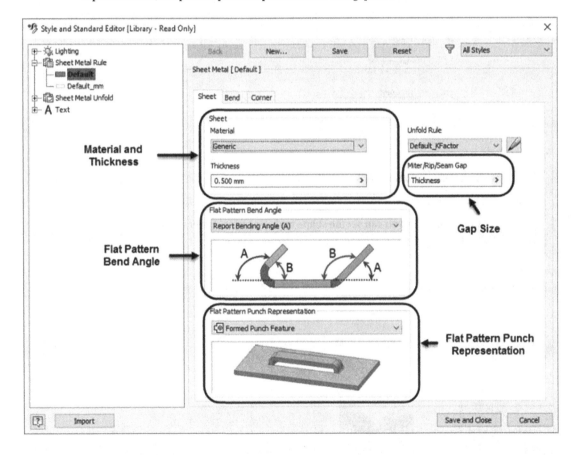

3. In the **Sheet** tab, set **Thickness** to 0.12 and leave all the default settings.

4. Click the **Bend** tab.

   In the **Bend** tab of this dialog, you can set the bend preferences such as bend radius, bend relief shape and size, and bend transition.

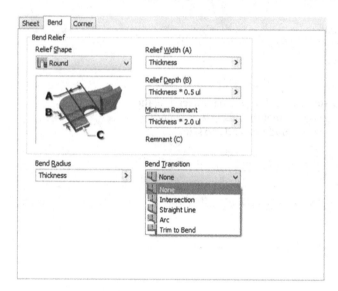

5. Set **Relief Shape** to **Round**.

6. Click the **Corner** tab.

   On the **Corner** tab, you can set the shape and size of the corner relief to be applied at the corners.

7. After setting the required preferences, click the **Save and Close** button.

The **Unfold Rule** option in the **Sheet Metal Defaults** dialog defines the folding/unfolding method of the sheet metal part. To modify or set a new unfold rule, click the **Edit Unfold Rule** button in the **Sheet Metal Defaults** dialog.

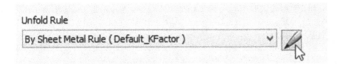

In the **Style and Standard Editor** dialog, select the required **unfold method**.

Unfold Method

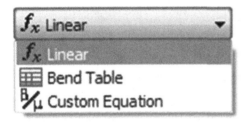

You can define the Unfold rule by selecting the **Linear** method (specifying the K factor), selecting a **bend table**, or entering a custom equation. Click **Save and Close** after setting the parameters.

8. Close the **Sheet Metal Defaults** dialog.

Creating the Base Feature

Follow these steps:

1. Create the sketch on the XZ plane, as shown here (use the **Rectangle Two Point Center** tool).

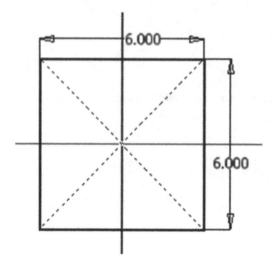

2. Click **Finish Sketch**.

3. To create the base component, click **Sheet Metal ➤ Create ➤ Face** on the ribbon; the **Face** dialog appears.

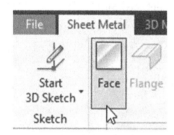

4. Click **OK** to create the tab feature.

Creating the Flange

Follow these steps:

1. To create the flange, click **Sheet Metal ➤ Create ➤ Flange** on the ribbon; the **Flange** dialog appears.

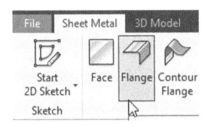

2. Select the edge on the top face, as shown here:

3. Set **Distance** to 4.

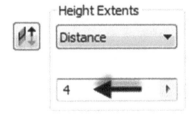

4. Click the **bend from the intersection of the two outer faces** icon in the **Height Datum** section. This measures the flange height from the outer face.

5. Under the **Bend Position** section, click the **Inside of the Bend Extents** icon.

6. Click **OK** to create the flange.

Creating the Contour Flange

Follow these steps:

1. Draw a sketch on the front face of the flange, as shown here:

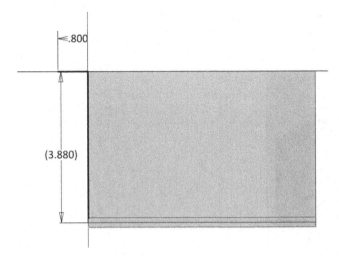

2. Click **Finish Sketch**.

3. To create the contour flange, click **Sheet Metal ➤ Create ➤ Contour Flange** on the ribbon; the **Contour Flange** dialog appears.

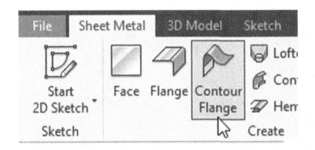

4. Select the sketch from the model.

5. Select the edge on the left side of the top face; the contour flange preview appears.

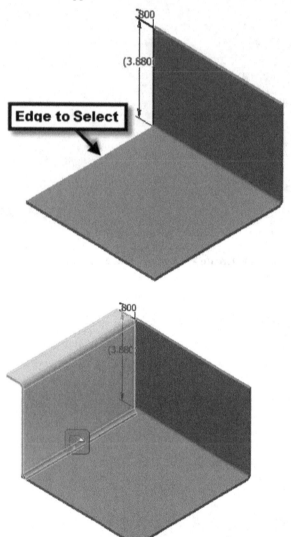

6. Select **Edge** from the **Type** drop-down.

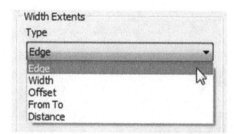

7. Click **OK** to create the contour flange.

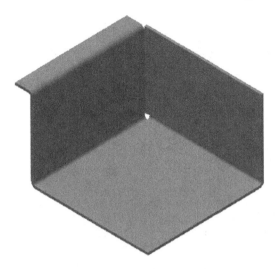

Creating the Corner Seam

Follow these steps:

1. To create the corner seam, click **Sheet Metal ➤ Modify ➤ Corner Seam** on the ribbon; the **Corner Seam** dialog appears.

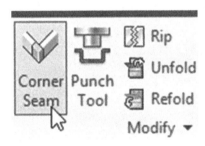

2. Rotate the model.

3. Select the two edges forming the corner.

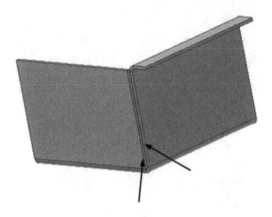

4. Set the parameters in the **Shape** tab of the dialog, as shown here:

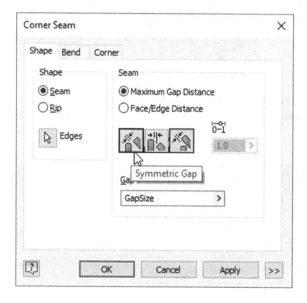

5. Click the **Bend** tab and make sure that the **Default** option is selected in the **Bend Transition** drop-down.

6. Click the **Corner** tab and set **Relief Shape** to **Round**.

   You can also apply other types of relief using the options in the **Relief Shape** drop-down.

7. Click **OK**.

Creating a Sheet Metal Punch iFeature

Follow these steps:

1. Open a new sheet metal file using the **Sheet Metal.ipt** template.

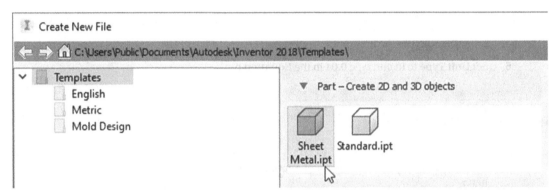

2. Create a sheet metal face of the dimensions 4×4.

3. Click **Manage ➤ Parameters ➤ Parameters** $f\!x$ on the ribbon; the **Parameters** dialog appears.

4. Select the **User Parameters** row and click the **Add Numeric** button in the dialog. This adds a new row.

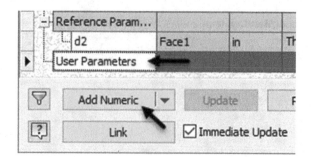

5. Enter **Diameter** in the new row.

6. Set **Unit Type** to **in** and type **0.04** in the **Equation** box.

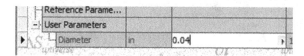

7. Likewise, create a parameter named **Length** and specify its values, as shown here:

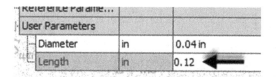

8. Click **Done**.

9. Click **Sheet Metal ➤ Sketch ➤ Start 2D Sketch** on the ribbon.

10. Select the top face of the base feature.

11. On the ribbon, click **Sketch ➤ Create**, select the **Rectangle drop-down**, and click **Slot Center to Center**.

12. Click to define the first center point of the slot.

13. Move the cursor horizontally and click to define the second center point of the slot.

14. Move the cursor outward and click to define the slot radius.

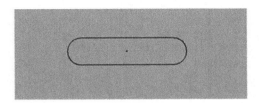

15. Click **Dimension** on the **Constrain** panel and select the round end of the slot.

16. Click to display the **Edit Dimension** box.

17. Click the arrow button on the box and select **List Parameters** from the shortcut menu; the **Parameters** list appears.

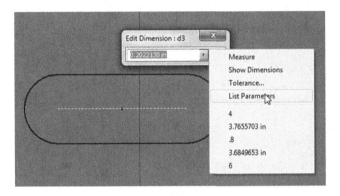

18. Select **Diameter** from the list and click the green check in the **Edit Dimension** box.

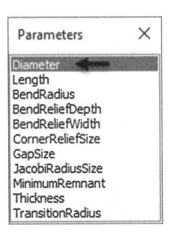

19. Likewise, dimension the horizontal line of the slot and set the parameter to **Length**.

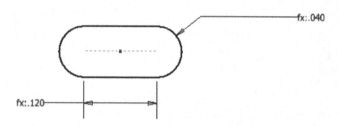

20. Click the **Point** button on the **Create** panel and place it at the center of the slot.

21. Delete any projected edges (yellow lines) from the sketch.

22. Click **Finish Sketch**.

23. Click **Sheet Metal ➤ Modify ➤ Cut** on the ribbon; the **Cut** dialog appears.

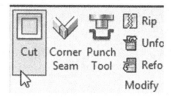

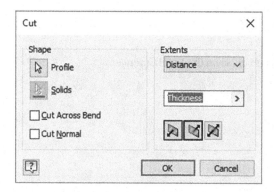

24. Accept the default values and click **OK** to create the cut feature.

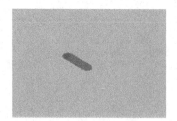

25. Click **Manage ➤ Author ➤ Extract iFeature** on the ribbon; the **Extract iFeature** dialog appears.

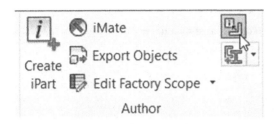

26. In the dialog, select **Type ➤ Sheet Metal Punch iFeature**.

27. Select the cut feature from the model geometry or from the Browser window. The parameters of the cut feature appear in the **Extract iFeature** dialog.

    Next, you must set the **size parameters** of the iFeature.

28. Set the **Limit** field of the **Diameter** value to **Range**. The **Specify Range** dialog appears.

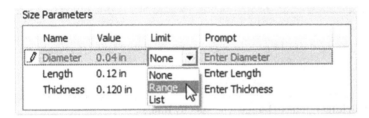

29. Set the values in the **Specify Range: Diameter dialog**, as shown here, and click **OK**.

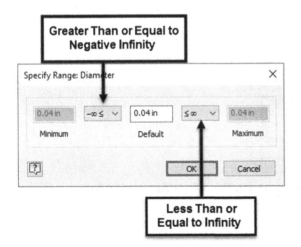

30. Set the **Limit** field of the **Length** value to **List**. The **List Values: Length** dialog appears.

31. Click **"Click here" to add a value** and enter **0.2** as the value.

32. Likewise, type values in the **List Values: Length** dialog, as shown here:

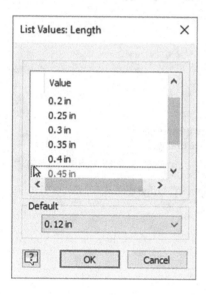

33. Click **OK**.

34. Set the **Limit** field of the **Thickness** value to **Range**. The **Specify Range: Thickness** dialog appears.

35. Set the values in the **Specify Range: Thickness** dialog, as shown here. Next, click **OK**.

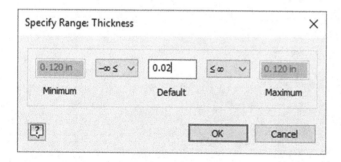

Next, you need to select the center point of the slot. This point will be used while placing the slot.

36. Click the **Select Sketch** button in the **Extract iFeature** dialog.

37. Select the sketch of the cut feature from the Browser window.

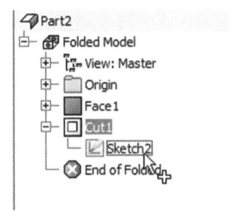

38. Click **Save** in the dialog; the **Save As** dialog appears.

39. Browse to the **Punches** folder and enter **Custom slot** in the **File name** box.

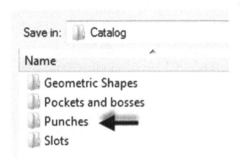

40. Click **Save** and **Yes**.

41. Click **File ➤ Save**.

42. Save the sheet metal part file as Custom slot.

43. Switch to the sheet metal file of the current tutorial.

Creating a Punched Feature

Follow these steps:

1. Start a sketch on the top face of the base sheet.

2. On the ribbon, click **Sketch ➤ Create ➤ Point**.

3. Place a point and add dimensions to it, as shown here:

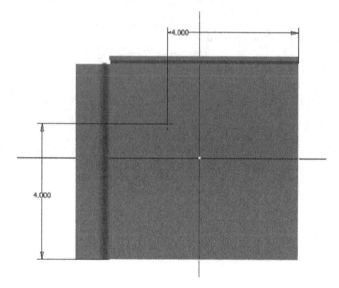

4. Click **Finish Sketch**.

5. To create the punch, click **Sheet Metal ➤ Modify ➤ Punch Tool** on the ribbon; the **PunchTool Directory** dialog appears.

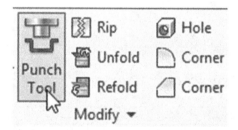

6. Select `Custom slot.ide` from the dialog and click **Open**; the **PunchTool** dialog appears.

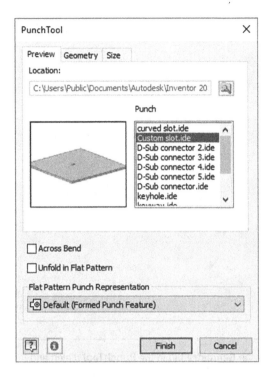

7. Click the **Size** tab in the **PunchTool** dialog.

8. Set **Length** to 0.45 and **Diameter** to 0.1.

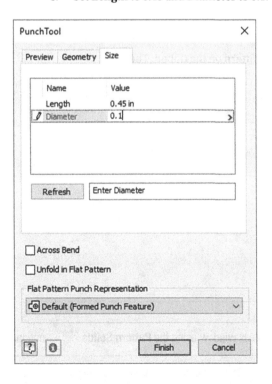

9. Click **Refresh** to preview the slot.

10. Click **Finish** to create the slot.

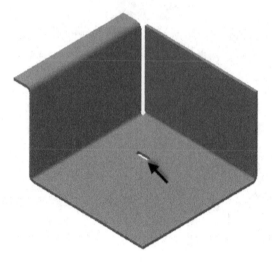

■ **Note**   If the slot is not oriented as shown here, then click the **Geometry** tab in the **PunchTool** dialog and type **90** in the **Angle** box.

Creating the Rectangular Pattern

Follow these steps:

1. Click **Sheet Metal ➤ Pattern ➤ Rectangular Pattern** on the ribbon. The **Rectangular Pattern** dialog appears.

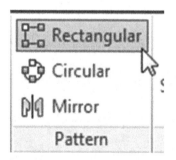

2. Select the slot feature.

 You can also select multiple solid bodies from the graphics window using the **Pattern Solids** option.

3.   Click the **Direction 1** button in the dialog.

4.   Select the edge of the base feature, as shown here:

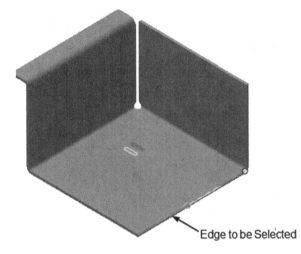

Edge to be Selected

5.   Select **Spacing** from the drop-down located in the **Direction 1** group.

6.   Specify **Column Count** as 5.

7.   Specify **Column Span** as 0.6.

8. Click the **Direction 2** button in the dialog.

9. Select the edge on the base feature, as shown here:

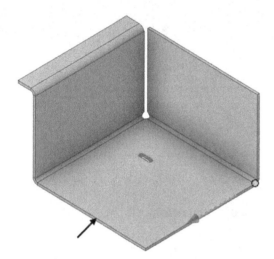

10. Click the **Flip** button in the **Direction 2** section to make sure the arrow is pointed toward the right.

11. Select **Spacing** from the drop-down located in the **Direction 2** group.

12. Specify **Column Count** as **2**.

13. Specify **Column Span** as **2**.

14. Click **OK** to create the pattern.

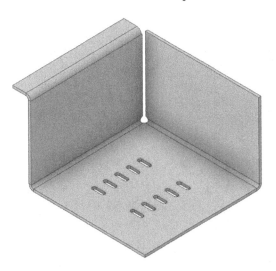

Creating the Bend Feature

Follow these steps:

1. Create a plane parallel to the front face of the flange feature. The offset distance is 6.3.

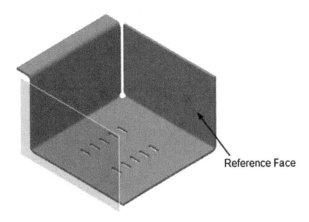

Reference Face

2. Create a sketch on the new workplane.

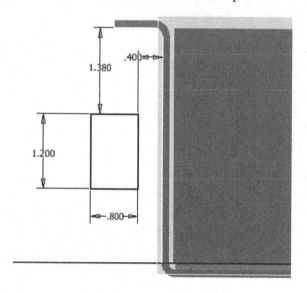

3. Click **Finish Sketch**.

4. Click **Sheet Metal ➤ Create ➤ Face** on the ribbon and create a face feature.

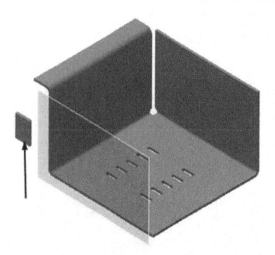

5. Click **Sheet Metal ➤ Create ➤ Bend** on the ribbon. The **Bend** dialog appears.

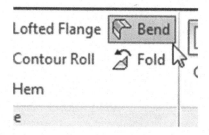

6. Select the edges from the model, as shown here:

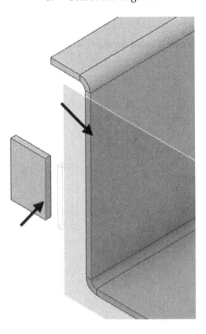

7. Make sure the **Bend Extension** is set to perpendicular.

8. Click **OK** to create the bend feature.

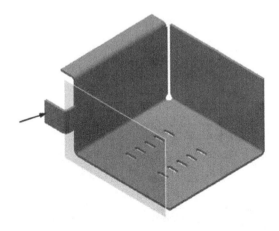

9. Hide the work plane (right-click it and deselect **Visibility**).

Applying a Corner Round

Follow these steps:

1. To apply a corner round, click **Sheet Metal ➤ Modify ➤ Corner Round** on the ribbon; the **Corner Round** dialog appears.

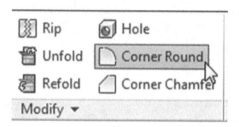

2. Set the **Radius** value to 0.2.

3. Set **Select Mode** to **Feature**.

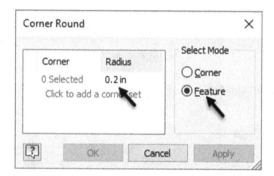

4. Select the face feature from the model.

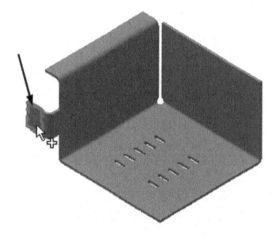

5. Click **OK** to apply the rounds.

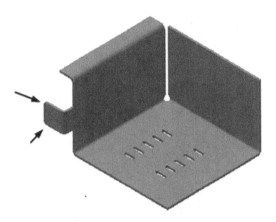

Creating Countersink Holes

Follow these steps:

1. Click **Sheet Metal ➤ Modify ➤ Hole** on the ribbon; the **Hole** dialog appears.

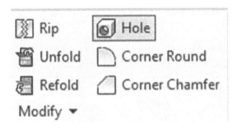

2. Set the **Placement** method to **Concentric**.

3. Set the hole type to **Countersink**.

4. Set the other parameters on the dialog, as shown here:

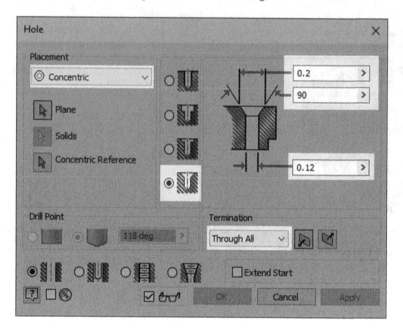

5. Click the face of the flange, as shown here:

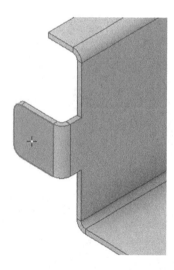

6. Select the corner round as the concentric reference.

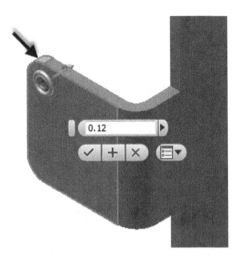

7. Click **Apply**.

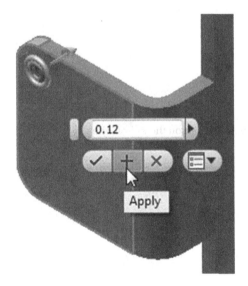

8. Again, click the flange face and select the other rounded corner as the concentric reference.

9. Click **OK** to create the countersink.

Creating Hem Features

Follow these steps:

1. To create the hem feature, click **Sheet Metal ➤ Create ➤ Hem** on the ribbon; the **Hem** dialog appears.

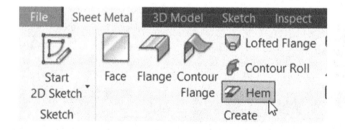

2. Set **Type** to **Single**.

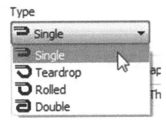

3. Select the edge of the contour flange, as shown here:

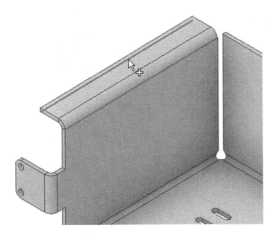

4. Leave the default settings of the dialog and click **OK** to create the hem.

Mirroring the Features

Follow these steps:

1. Click **Mirror** 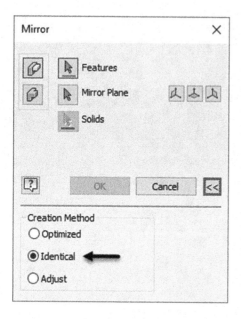 on the **Pattern** panel; the **Mirror** dialog appears.

2. Click ➤ at the bottom of the dialog and make sure **Creation Method** is set to **Identical**.

3. Select the features from the Browser window, as shown here:

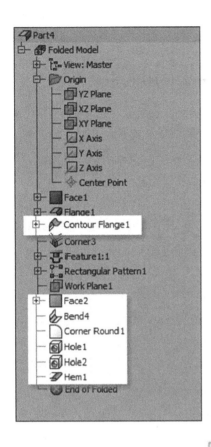

4. Click the **Origin YZ Plane** 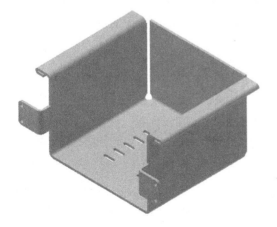 button in the dialog.

5. Click **OK** to mirror the feature.

6. Create a corner seam between the mirrored counter flange and flange.

Creating the Flat Pattern

Follow these steps:

1. To create a flat pattern, click **Sheet Metal ➤ Flat Pattern ➤ Create Flat Pattern** on the ribbon.

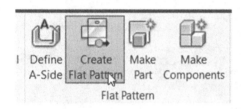

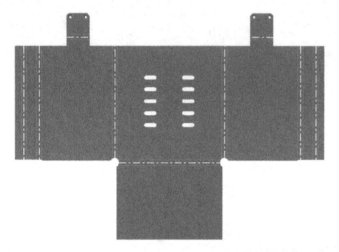

You can set the order in which the bends will be annotated.

2. Click the **Bend Order Annotation** button on the **Manage** panel of the **Flat Pattern** tab. The order in which the bends will be annotated is displayed.

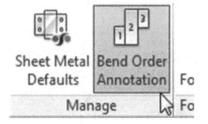

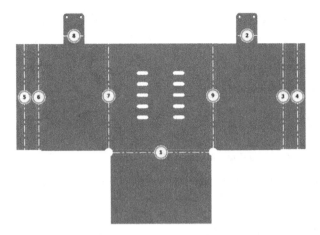

3. To change the order of the bend annotation, click the balloon displayed on the bend. The **Bend Order Edit** dialog appears.

4. Select the **Bend Number** check box and enter a new number in the data field.

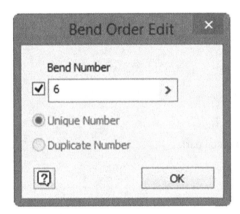

5. Click **OK** to change the order.

6. To switch back to the folded view of the model, click **Go to Folded Part** on the **Folded Part** panel.

7. Save the sheet metal part.

301

Creating 2D Drawing of the Sheet Metal Part

Follow these steps:

1. On the Quick Access Toolbar, click the **New** button.

2. In the **Create New File** dialog, double-click **Standard.idw**.

3. Activate the **Base View** tool.

4. Click Home icon on the ViewCube.

5. Leave the default settings on the Drawing View dialog and click **OK**.

6. Click and drag the drawing view to the top-right corner of the drawing sheet.

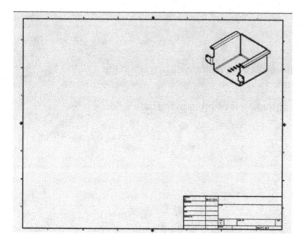

7. Likewise, create the front and top views of the sheet metal part.

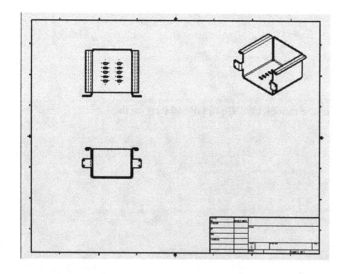

8. Activate the **Base View** tool and select **Sheet Metal View ➤ Flat Pattern** in the **Drawing View** dialog.

9. Place the flat pattern view below the Isometric view.

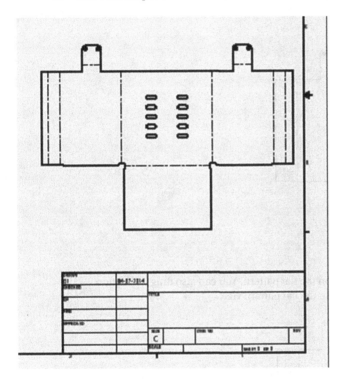

10. To add bend notes to the flat pattern, click **Annotate ➤ Feature Notes ➤ Bend** on the ribbon.

11. Click the horizontal bend line on the flat pattern to add the bend note.

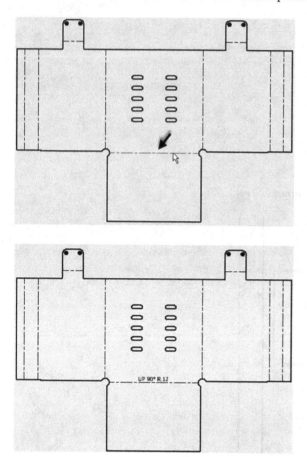

12. Likewise, select other bend lines on the flat pattern. You can also drag a selection box to select all the bend lines from the flat pattern view.

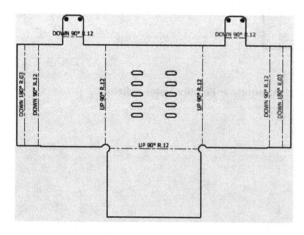

13. To add centerlines to the flat pattern view, right-click and select **Automated Centerlines**.

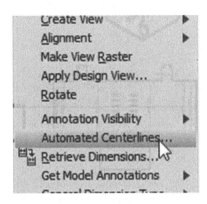

14. In the **Automated Centerlines** dialog, click the **Punches** button under the **Apply To** section.

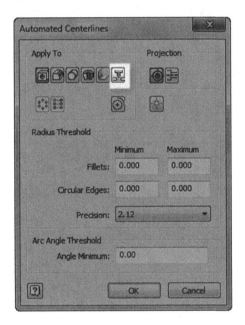

15. Click **OK** to add centerlines to the flat pattern view.

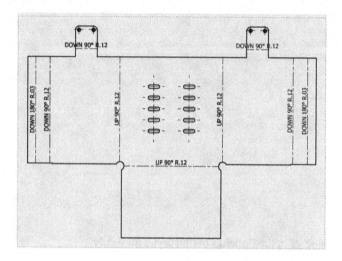

16. Likewise, add centerlines to other views on the drawing sheet.

17. To add a punch note, click **Annotate ➤ Feature Notes ➤ Punch** on the ribbon.

18. Zoom into the flat pattern view and click the arc of the slot.

19. Move the pointer and click to create the annotation.

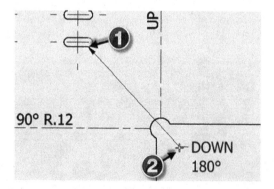

20. Use the **Retrieve Dimension** and **Dimension** tools to add dimensions to drawing.

21. Save and close the drawing and sheet metal part.

CHAPTER 7

■ ■ ■

Top-Down Assembly and Joints

In this chapter, you will learn to do the following:

- Create a top-down assembly
- Insert fasteners using Design Accelerator
- Export to 3D PDF
- Create assembly joints

Tutorial 1

In this tutorial, you will create the model shown here. You will use the top-down assembly approach to create this model.

© T. Kishore 2017
T. Kishore, *Learn Autodesk Inventor 2018 Basics*, https://doi.org/10.1007/978-1-4842-3225-5_7

84

Creating a New Assembly File

Follow these steps:

1. To create a new assembly, click **New Assembly** on the Home screen.

Creating a Component in the Assembly

In a top-down assembly approach, you create components of an assembly directly in the assembly by using the **Create** tool.

1. Click **Create** on the **Component** panel of the **Assembly** tab. The **Create In-Place Component** dialog appears.

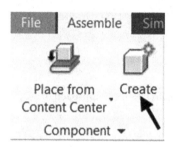

2. Enter **Base** in the **New Component Name** field.

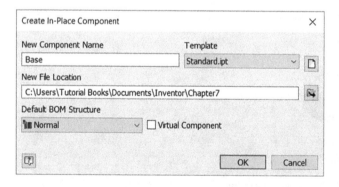

3. In the **Create In-Place Component** dialog, set **New File Location** to the current project folder.

4. Click the **Browse to New File Location** icon.

5. In the **Save As** dialog, click the **Create New Folder** icon.

6. Type **C07_Tut_01** as the name of the folder.

7. Double-click the new folder and click **Save**.

8. Click **OK** in the **Create In-Place Component** dialog.

9. Expand the **Origin** folder in **Browser window** and select **XZ Plane**. The **3D Model** tab is activated in the ribbon.

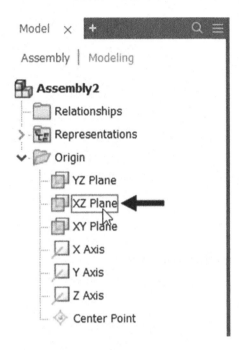

10. Click **Sketch ➤ Start 2D Sketch** on the ribbon.

11. Select **XZ Plane**.

12.   Create a sketch as shown here:

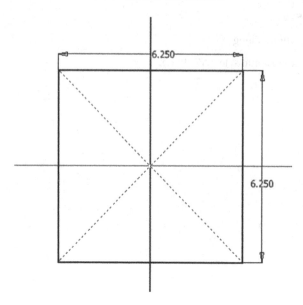

13.   Click **Finish Sketch**.

14.   Click **3D Model ➤ Create ➤ Extrude** on the ribbon and extrude the sketch up to 1.5 in.

15. Start a sketch on the top face and draw a circle that has a diameter of **2** in.

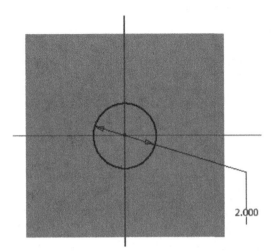

16. Click **Finish Sketch**.

17. Extrude the sketch up to a distance of 3.75 in.

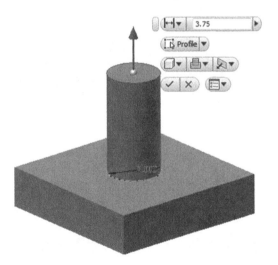

18. Create a counterbore hole on the second feature (see Chapter 5's Tutorial 1).
The following figure shows the dimensions of the counterbore hole:

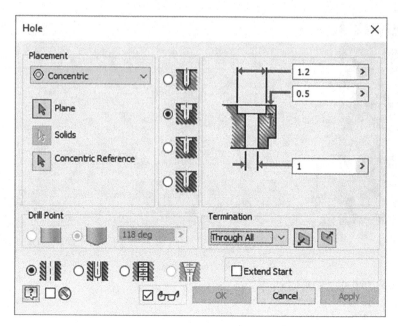

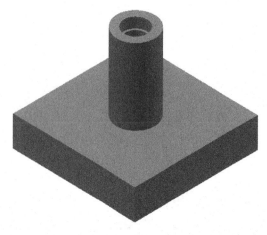

19. Start a new sketch on the top face of the first feature.

20. Create a 3.5-diameter circle with the **Construction** button active.

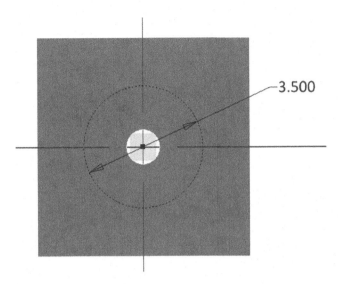

21. On the ribbon, click **Sketch ➤ Create ➤ Point**.

22. Place a point on the circle, as shown here:

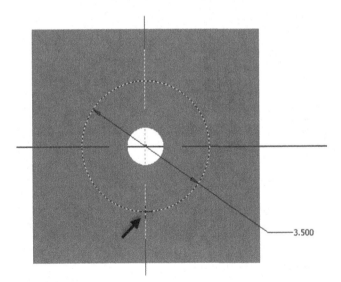

23. Click **Finish Sketch**.

24. On the ribbon, click **3D Model ➤ Modify ➤ Hole**.

25.   In the **Hole** dialog, specify the settings as shown here:

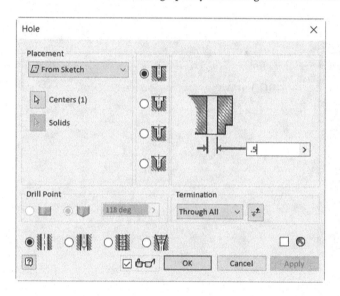

26.   Make sure that the sketch point is selected.

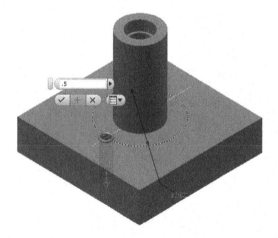

27.   Click **OK** to create the hole.

28.   Create a circular pattern of the hole (see Chapter 5's Tutorial 1).

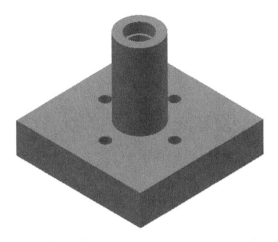

29.   Click the **Return** button on the ribbon.

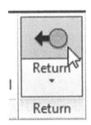

Creating the Second Component of the Assembly

Follow these steps:

1.   Click **Assemble ➤ Component ➤ Create** 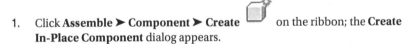 on the ribbon; the **Create In-Place Component** dialog appears.

2.   Enter **Spacer** in the **New Component Name** field.

3.   Select the "**Constrain sketch plane to selected face or plane**" option.

4.   Click **OK**.

315

5. Select the top face of the base.

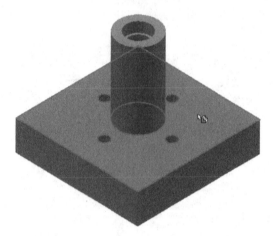

6. Click **Sketch ➤ Start 2D Sketch** on the ribbon.

7. Select the top face of the base.

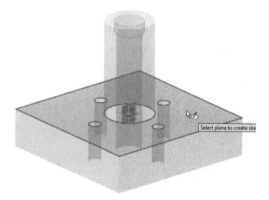

8. On the ribbon, click **Sketch ➤ Create ➤ Project Geometry** and select the circular edges of the base.

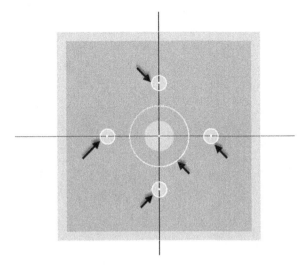

9.   Draw a circle with a diameter of 4.5 in.

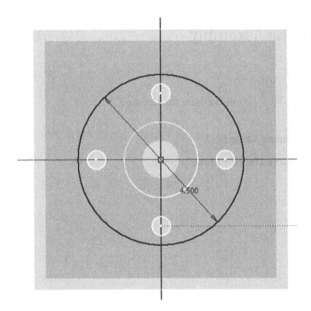

10.   Click **Finish Sketch**.

11.   Extrude the sketch up to 1.5 in.

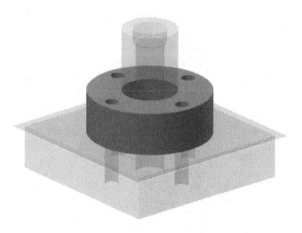

12.   Click **Return**  on the ribbon.

Creating the Third Component of the Assembly

Follow these steps:

1.   Click **Assemble ➤ Component ➤ Create** on the ribbon; the **Create In-Place Component** dialog appears.

2.   Enter **Shoulder Screw** in the **New Component Name** field.

3.   Select the "**Constrain sketch plane to selected face or plane**" option.

4.   Click **OK**.

5.   Click the top face of the base.

6.   Start a sketch on the YZ plane.

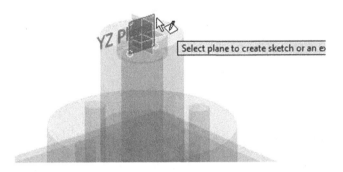

7.   Draw a sketch, as shown here:

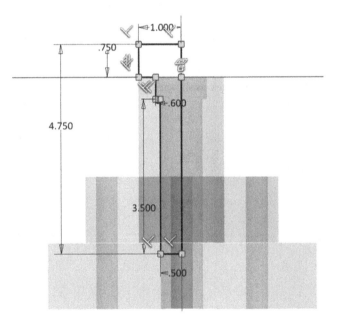

8.   Click **Finish Sketch**.

9. Activate the **Revolve** tool and revolve the sketch.

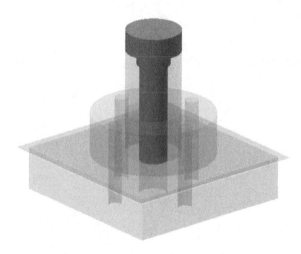

10. Activate the **Chamfer** tool and chamfer the edges, as shown here:

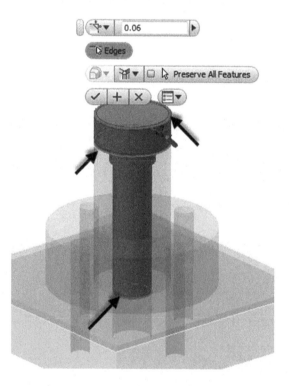

11. Activate the **Fillet** tool and round the edges, as shown here:

12. Click **Return** 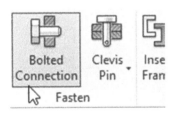 on the ribbon.

13. Save the assembly.

Adding Bolt Connections to the Assembly

Follow these steps:

1. On the ribbon, click **Design ➤ Fasten ➤ Bolt Connection**.

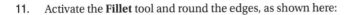

2. In the **Bolted Connection Component Generator** dialog, on the **Design** tab, select **Type ➤ Through All**.

3.   Select **Placement ➤ Concentric**.

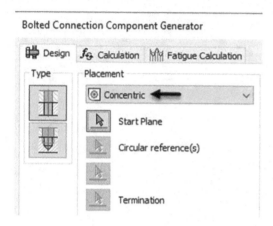

4.   Select the top face of the spacer.

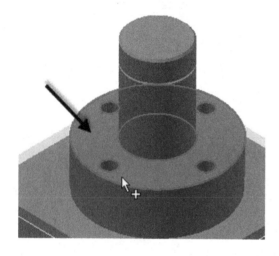

5.   Click the hole to define the circular reference.

6. Rotate the model and click the bottom face of the base. This defines the termination.

7. In the dialog, set the **thread** type to **ANSI Unified Screw Threads**.

8. Make sure that **Diameter** is set to **0.5** in.

9. In the dialog, click "**Click to add a fastener**."

10.   In the pop-up dialog, set **Standard** to **ANSI** and **Category** to **Hex Head Bolt**.

11.   Select **Hex Bolt-Inch**. This adds a hex bolt to the list.

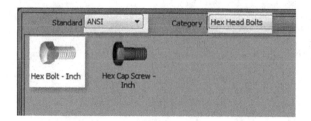

12.   In the list, click **"Click to add a fastener"** below the hex bolt.

13.   In the pop-up dialog, scroll down and select **Plain Washer (Inch)**.

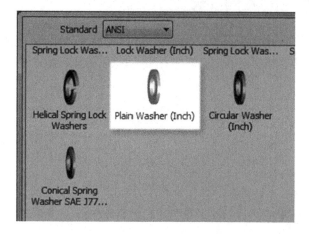

14.   Click **"Click to add a fastener"** at the bottom of the list.

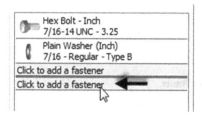

15.   In the pop-up dialog, scroll down and select **Plain Washer (Inch)**.

16.   Click **"Click to add a fastener"** at the bottom of the list.

17.   In the pop-up dialog, set **Category** to **Nuts** and select **Hex Nut – Inch**.

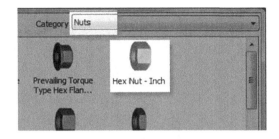

18. Click **OK** twice to add a bolt connection subassembly.

Patterning Components in an Assembly

Follow these steps:

1. On the ribbon, click **Assemble ➤ Pattern ➤ Pattern**.

2.   Select the **Bolt connection** from the Browser window.

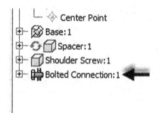

3.   In the **Pattern Component** dialog, click the Circular tab and select the Axis Direction button.

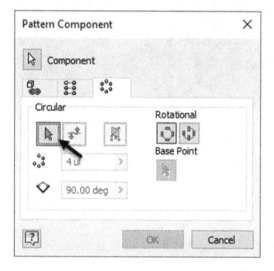

4.   Click the large cylindrical face of the spacer to define the axis of the circular pattern.

5.   In the dialog, type **4** and **90** in the **Circular Count** and **Circular Angle** boxes, respectively.

6.   Click **OK** to pattern the bolt connection.

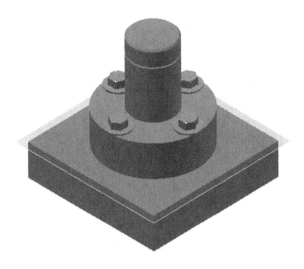

Applying the Constraint to the Components

Follow these steps:

1. On the ribbon, click **View** ➤ **Visibility** ➤ **Degrees of Freedom**.

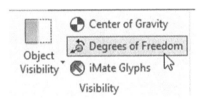

2. On the ribbon, click **Assemble ➤ Relationships ➤ Constrain**.

3. In the dialog, click the **Mate** icon and click the cylindrical faces of the spacer and base.

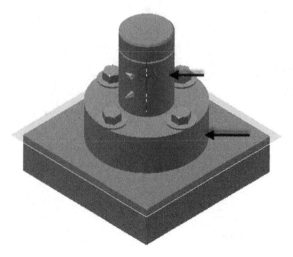

4. Click **Apply**.

5. Click the cylindrical faces of the shoulder screw and base.

6. Click **Apply**.

7. In the dialog, select **Flush** from the **Solution** section.

8. In the Browser window, expand the **Origin** folder and select XY Plane.

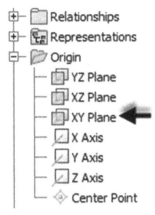

9.   Expand the **Origin** folder of the shoulder screw and select XZ Plane.

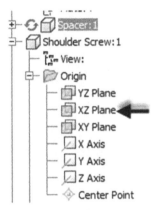

10.   Click **OK** to fully constrain the assembly.

11.   Save the assembly and all its parts.

Using the Search tool in the Browser Window

Autodesk Inventor 2018 provides you with the search tool to locate components and features quickly.

1.   In the Browser window, click the **Search** 🔍 icon.

2. Type **hex** in the search bar; all the hexagonal bolts appear in the Browser window.

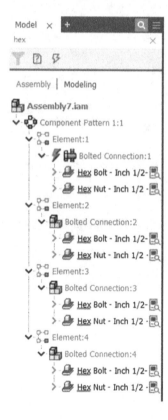

3. Place the pointer on the hexagonal bolts in the Browser window; they are highlighted in the graphics window.

    You can select all the hexagonal bolts by pressing the Shift key and clicking them. After selecting them, you can perform a variety of operations at a time such as hiding, deleting, solving, suppressing, and so on.

4. Click **Clear Search** to clear all the searched components.

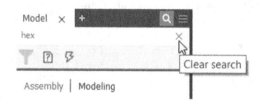

Editing Values in the Browser Window

Autodesk Inventor 2018 allows you to edit the values of the assembly components directly in the Browser window.

1. In the Browser window, click the Drop-down menu next to the Search box and then select **Edit Values in Browser**.

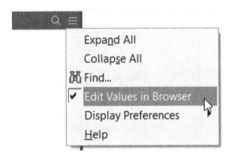

2. In the Browser window, expand the Shoulder Screw part and then click the Flush relation, as shown; the selected relation is highlighted in the graphics window, as shown here:

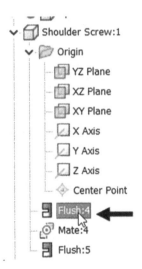

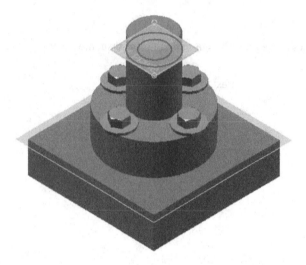

3. Type **1** in the box that appears next to the selected relation and then press Enter;
    the relation is updated in the graphics window.

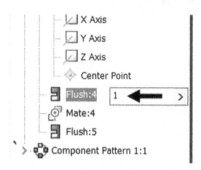

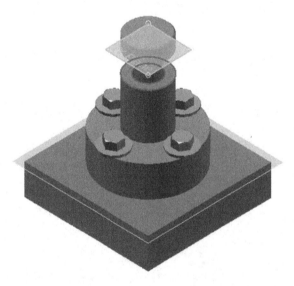

4.   Click **Undo** on the Quick Access Toolbar.

Changing the Display Preferences of the Browser Window

Autodesk Inventor 2018 allows you to hide or display items to reduce the clutter in your Browser window. For example, you can hide or display the work features such as the planes and UCS in the Browser window.

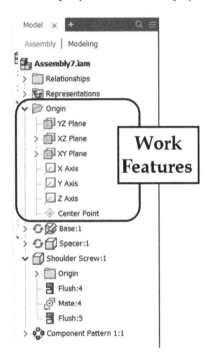

1. In the Browser window, click the drop-down menu next to the **Search** box and then select **Display Preferences ➤ Hide Work Features**.

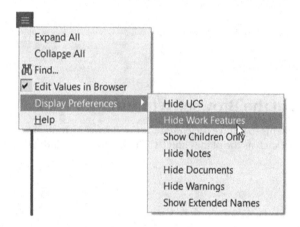

The work features are hidden.

Using the Measure Tool

The **Measure** tool helps you measure the size and position of the model. You can measure the various parameters of the model such as length, angle, radius, and so on.

1. On the ribbon, click **Inspect ➤ Measure ➤ Measure** ; the **Measure** floating window appears on the screen.

2. Click and drag the **Measure** floating window and then release it in the Browser window; the Measure window is docked to the Browser window.

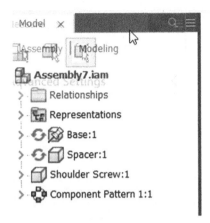

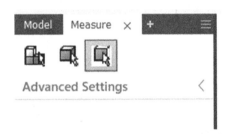

The **Measure** window has three selection filters (shown from right to left): **Select Faces and Edges, Part Priority**, and **Component Priority**.

The **Select Faces and Edges** filter allows you to select only the faces and edges of the model.

The **Part Priority** filter allows you to select the part geometry for measurement.

The **Component Priority** filter allows you select the part geometry and assemblies. This filter is used to select subassemblies from a main assembly.

3. Select the **Select Faces and Edges**  filter and select the linear edge, as shown here:

The length of the selected edge is displayed in the **Measure** window.

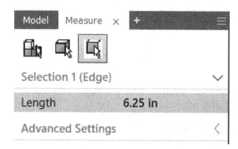

4. Click **Advanced Settings** in the **Measure** window.

In the **Advanced Settings** section, you can change the **Precision** and **Angle Precision** settings of the displayed measurement. In addition, you can display the measurement in dual units by specifying the **Dual Units** type.

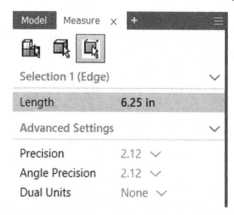

5.   Select the cylindrical face, as shown; the Measure window displays results.

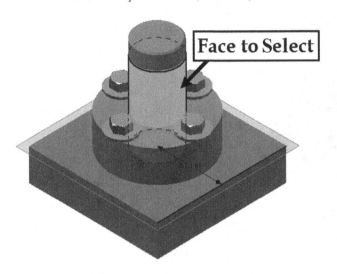

The "Measure results" section displays the results of the first and second selections separately. In addition to that, the distance between the two selected entities is displayed.

6.   Save and close the assembly and its parts.

Tutorial 2

In this tutorial, you create a slider crank mechanism by applying joints.

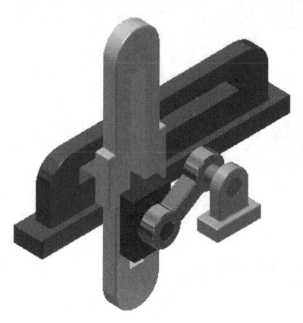

1. Create the **Slider Crank Assembly** folder inside the project folder.

2. Download the part files of the assembly from the companion web site. Next, save the files in the **Slider Crank Assembly** folder.

3. Start a new assembly file using the **Standard.iam** template.

4. Click **Assemble ➤ Component ➤ Place** on the ribbon.

5. Browse to the **Slider Crank Assembly** folder and double-click **Base**.

6. Right-click and select **Place Grounded at Origin**.

7. Right-click and select **OK**.

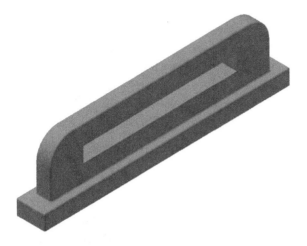

8. Click **Assemble ➤ Component ➤ Place** on the ribbon.

9. Browse to the **Slider Crank Assembly** folder and select all the parts except the **base**.

10. Click **Open** and click in the graphics window to place the parts.

11. Right-click and select **OK**.

12. Click and drag the parts, if they are coinciding with each other.

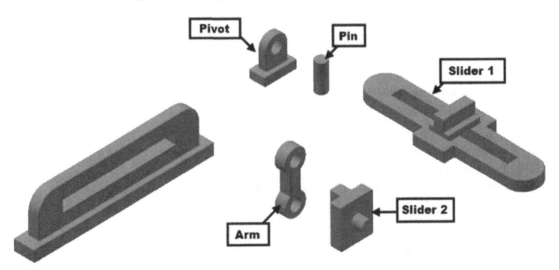

Creating the Slider Joint

Follow these steps:

1. Click **Assemble ➤ Relationships ➤ Joint** on the ribbon; the **Place Joint** dialog appears.

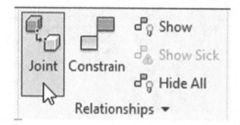

2. Set **Type** to **Slider**.

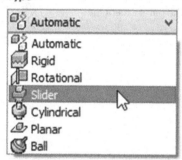

3. Select the face on the Slider1, as shown here:

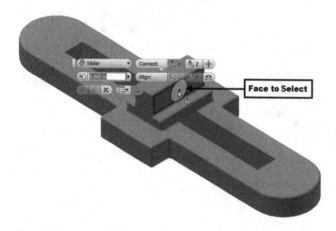

4. Select the face on the base, as shown here; the two faces are aligned.

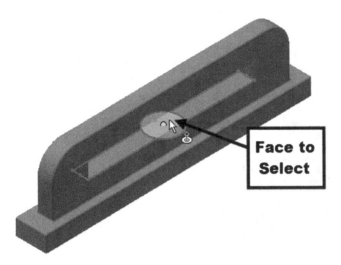

5. In the dialog, click the **First Alignment** button.

6. Select the face of Slider1, as shown here:

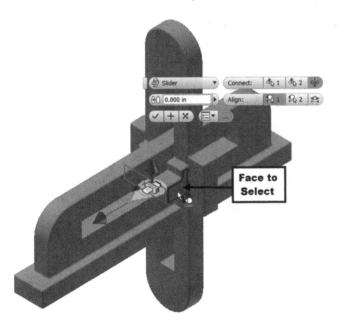

7.   Select the face of the base, as shown; Slider 1 is aligned to the selected face.

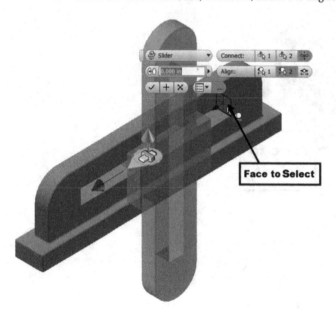

8.   Click the **Limits** tab in the **Place Joint** dialog.

9.   Check the **Start** and **End** options under the **Linear** group.

10. Set the **Start** value to 1.5 in and the **End** value to -1.5 in.

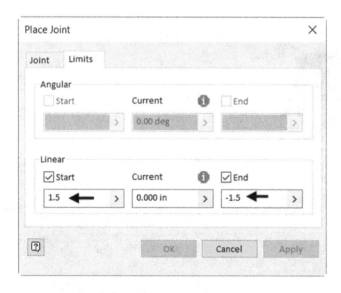

11. Click **OK**.

12. Select Slider1 and drag the pointer; Slider1 slides in the slot of the base. Also, the slider motion is limited up to the end of the slot.

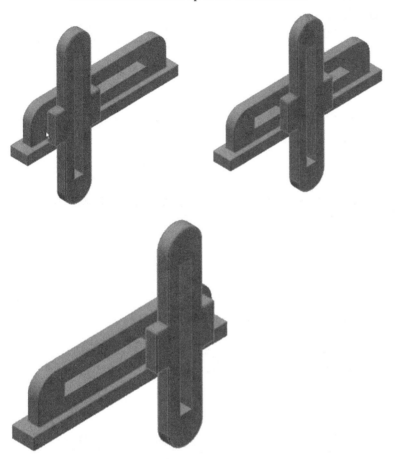

13. Click the corner of the ViewCube, as shown; the orientation of the assembly is changed.

14. Click **Assemble ➤ Relationships ➤ Joint** on the ribbon.

15.   In the dialog, set **Type** to **Slider** .

16.   Select the face on the Slider2, as shown here:

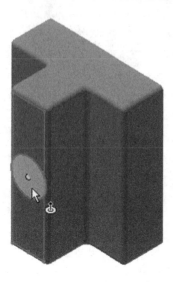

17.   Select the right edge of the top face of the ViewCube; the orientation of the assembly changes.

18.   Select the face on Slider1, as shown here. Next, click **OK**.

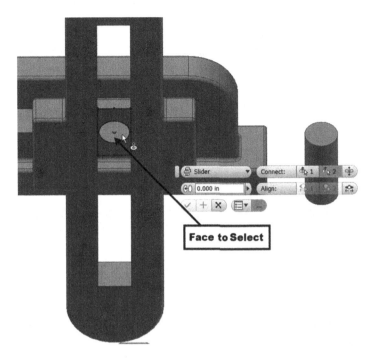

Creating the Rotational Joint

Follow these steps:

1.   Click **Assemble ➤ Relationships ➤ Joint** on the ribbon.

2.   Set **Type** to **Rotational**.

Type

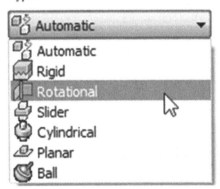

3. Select the circular edge of the arm, as shown here:

4. Select the circular edge of Slider2.

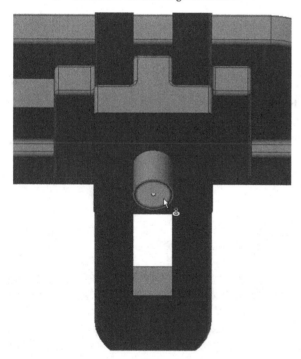

5. Click the **Flip Component** button under the **Connect** group.

6. Click **OK**.

Creating the Rigid Joint

Follow these steps:

1. Click **Assemble ➤ Relationships ➤ Joint** 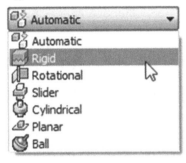 on the ribbon.

2. Set **Type** to **Rigid**.

3. Select the top face on the pin.

4. Click the corner point of the ViewCube, as shown here:

5. Select the circular edge on the back face of the arm.

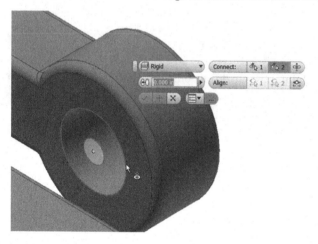

6. Click **OK**.

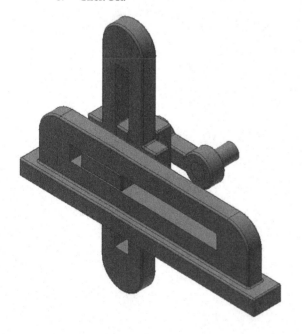

Adding More Assembly Joints

Follow these steps:

1. Create another rotational joint between the pin and the pivot.

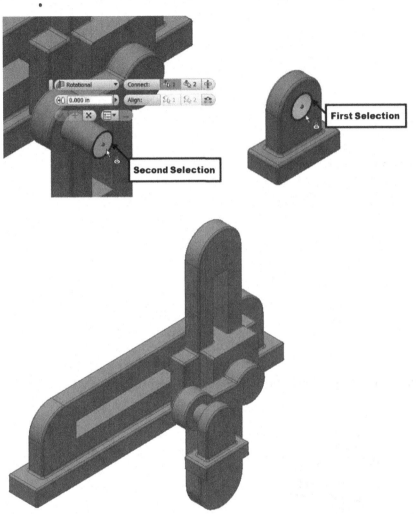

Next, you need to constrain the pivot by applying constraints.

2. Click the **Assemble** button on the **Relationships** panel.

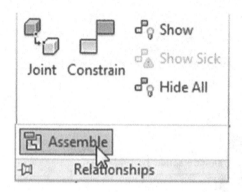

3. On the **Assembly** mini-toolbar, select **Mate – Flush** from the drop-down.

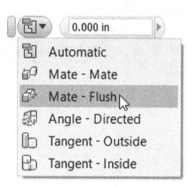

4. Select the bottom face of the pivot and then select the bottom face of the base.

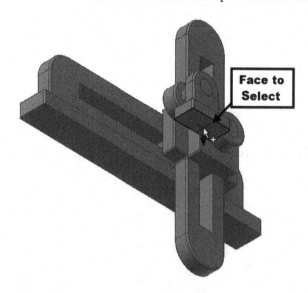

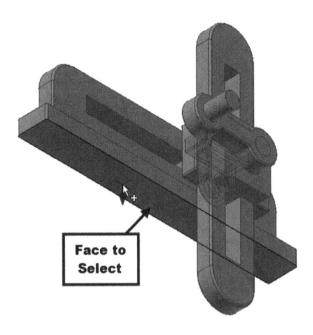

Face to
Select

5. Click **Apply** (plus symbol on the mini-toolbar).

6. Select the **XY plane** of the pivot and the **XY plane** of the base from the **Browser window**.

▲ 🗁 Pivot:1

   ▲ 🗀 Origin

      🔲 YZ Plane

      🔲 XZ Plane

      🔲 XY Plane ◀━━

      ▱ X Axis

      ▱ Y Axis

      ▱ Z Axis

     ◈ Center Point

   🗍 Rotational:2

   🗎 Flush:1

   🗎 Flush:2

- ▲ 🔯 Base:1
  - ▲ 🗁 Origin
    - 🗐 YZ Plane
    - 🗐 XZ Plane
    - 🗐 XY Plane ◀──
    - ⬚ X Axis
    - ⬚ Y Axis
    - ⬚ Z Axis
    - ◈ Center Point
  - 👤 Slider:1 +/-
  - 🗐 Flush:1
  - 🗐 Flush:2

7. Click **OK** (check mark on the mini-toolbar).

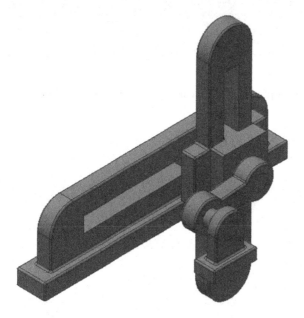

Driving the Joints

Follow these steps:

1. In the Browser window, expand Pivot and right-click the **Rotational** joint.

2. Select **Drive** from the shortcut menu.

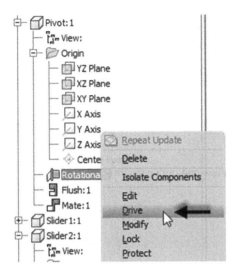

3. In the **Drive** dialog, type **0** and **360** in the **Start** and **End** boxes, respectively.

4. Expand the dialog by clicking the double-arrow button located at the bottom. In the expanded dialog, you can define the settings such as drive adaptivity, collision detection, increment, repetition, and so on.

5. Click the **Record** ⊙ button on the dialog. Specify the name and location of the video file. Click **Save** and **OK**.

6. In the dialog, click the **Forward** 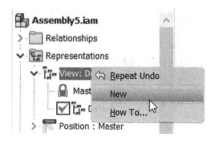 button to simulate the motion of the slider crank assembly.

7. Click **OK** to close the dialog.

Creating Positions

Follow these steps:

1. In the Browser window, expand **Representations ➤ View** and notice that the **Master** representation is set as the default.

2. Right-click the **Position** node and then select **New**; a new position is created.

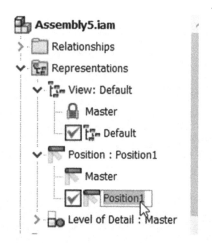

3. Double-click **Position1** and type **StartPosition**; the view representation is renamed.

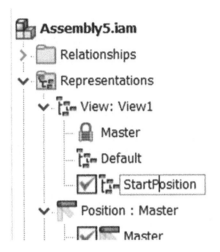

4. Click and drag Slider1 to the left end, as shown here:

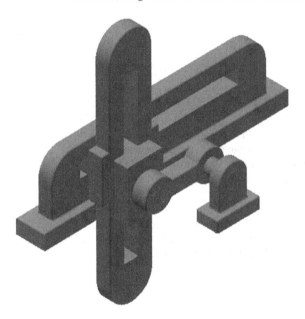

5. Double-click the **Master** positon to activate it.

Creating 3D PDF

Autodesk Inventor allows you to create a 3D PDF from the model. The 3D PDF file is helpful in viewing the 3D PDF without any CAD application or viewer.

1. Click **File ➤ Export ➤ 3D PDF**.

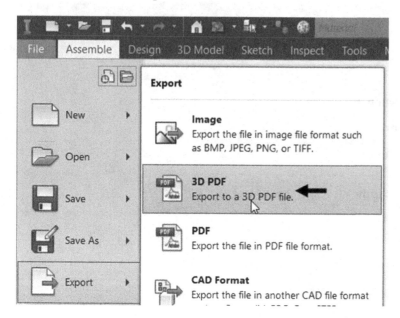

The **Publish 3D PDF** dialog appears on the screen. It is powered by **Anark Core** software. In the Publish 3D PDF dialog, you can select the properties to be displayed on the PDF from the **Properties** section. You can also select the required design view representation, visualization quality, and export scope.

2. Leave Template set to the default setting.

   If required, you can select another template by clicking the icon next to the Template path. You can also create a new 3D PDF template, if you have Adobe Acrobat Pro. You can go through the Autodesk Inventor Help file to know the procedure to create a 3D PDF template.

3. Specify the **File Output Location** setting.

4. Select the "**View PDF when finished**" option.

5. Select the "**Generate and attach STEP file**" option.

6. Click the **Options** button next to the "**Generate and attach the STEP file**" option; the **STEP file Save as Options** dialog appears on the screen.

   In this dialog, select the required Application Protocol option and spline fit accuracy. You can also enter the authorization, author, organization, and description.

7. Click **OK** in the "**STEP file save as Options**" dialog.

   Use the **Attachments** button, if you want to add any other attachments to the PDF file such as spreadsheet, PDF, or text document.

8. Click **Publish** in the **Publish 3D PDF** dialog.

Inventor starts exporting the 3D model to the PDF file. After a few seconds, the
3D PDF file opens in the PDF viewer.

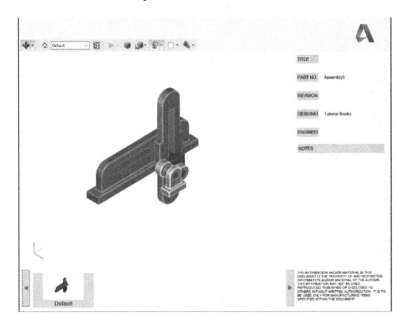

9. Click inside the graphics window of the PDF file and then drag to rotate the
model.

10. Click the drop-down located at the top-left corner and notice the **View** options.
These options are same as that available on the Navigation Bar.

357

Likewise, examine the other options on the toolbar. These options are similar to that available on the View ribbon tab of the Autodesk Inventor application.

11. On the sidebar, click the Attachments icon to view the STEP file.

You can open or save the STEP file by right-clicking it and selecting the corresponding option.

12. Close the 3D PDF file.

13. Save and close the assembly and its parts.

CHAPTER 8

■ ■ ■

Dimensions and Annotations

In this chapter, you will learn to do the following:

- Create centerlines and centered patterns
- Edit a hatch pattern
- Apply dimensions
- Place hole callouts
- Place leader text
- Place datum features
- Place feature control frames
- Place surface texture symbols
- Modify title block information

© T. Kishore 2017

T. Kishore, *Learn Autodesk Inventor 2018 Basics*, https://doi.org/10.1007/978-1-4842-3225-5_8

Tutorial 1

In this tutorial, you will create the drawing shown here:

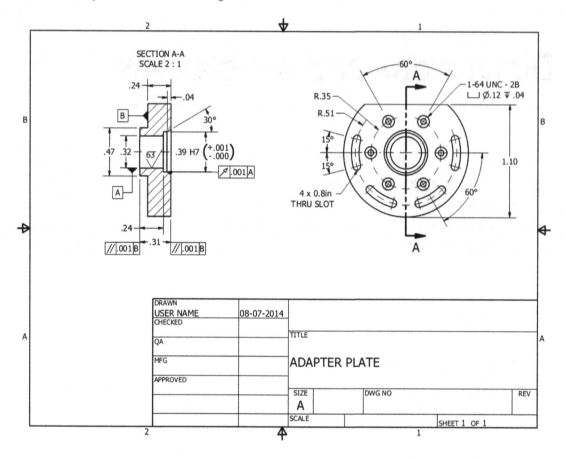

1. Open a new drawing file using the **Standard.idw** template.

2. In the Browser window, right-click Sheet:1 and select **Edit Sheet**.

3. In the **Edit Sheet** dialog, select **Size ➤ A** and then click OK.

4. Click **Place Views ➤ Create ➤ Base** on the ribbon.

5. Click **Open an Existing File** button in the dialog.

6. Browse to the location of the adapter plate created in Chapter 5's Tutorial 1. You can also download this file from the companion web site and use it.

7. Select the adapter plate file and click **Open**.

8. Set **Scale** to 2:1.

9. Click the front face on the ViewCube displayed in the drawing sheet.

10. Set **Style** to Hidden Line Removed .

11. Click OK in the dialog.

12. Drag the view to the right side of the drawing sheet.

13. Click **Place Views ➤ Create ➤ Section** 🔲 on the ribbon.

14. Select the front view.

15. Draw the section line on the front view.

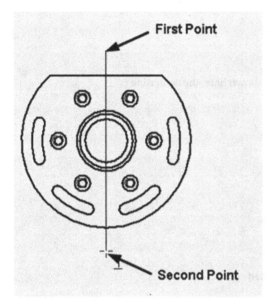

16. Right-click and select **Continue**.

17. Place the section view on the left side.

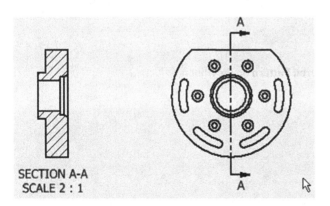

Creating Centerlines and Centered Patterns

Follow these steps:

1. Click **Annotate ➤ Symbols ➤ Centerline Bisector** on the ribbon.

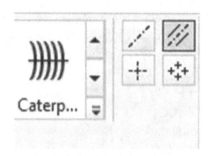

2. Select the parallel lines on the section view, as shown here; the centerline is created.

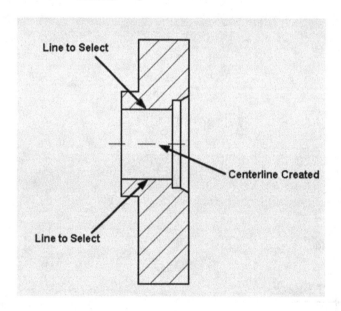

3. Click **Annotate ➤ Symbols ➤ Centered Pattern** on the ribbon.

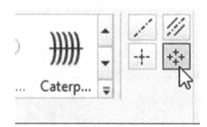

4.   Select the circle located at the center.

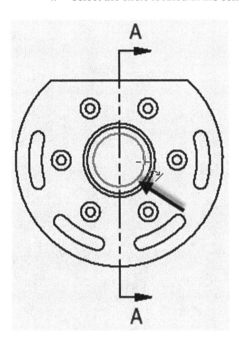

5.   Select the center point of any one of the counterbored holes.

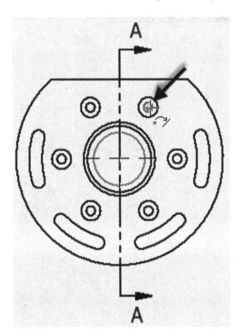

6. Select the center points of other counterbored holes.

7. Right-click and select **Create**.

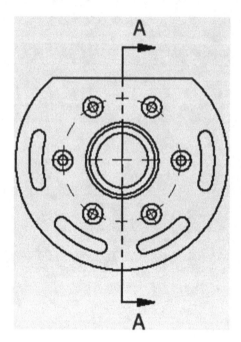

8. Likewise, create another centered pattern on the curved slots. Right-click and select **Create**.

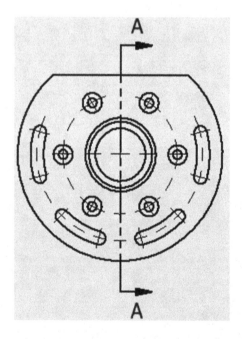

9. Press Esc to deactivate the tool.

Editing the Hatch Pattern

Follow these steps:

1. Double-click the hatch pattern of the section view; the **Edit Hatch Pattern** dialog appears.

2. You can select the required hatch pattern from the **Pattern** drop-down. If you select the **Other** option from this drop-down, the **Select Hatch Pattern** dialog appears. You can select a hatch pattern from this dialog or load a user-defined pattern by using the **Load** option. Click **OK** after selecting the required hatch pattern.

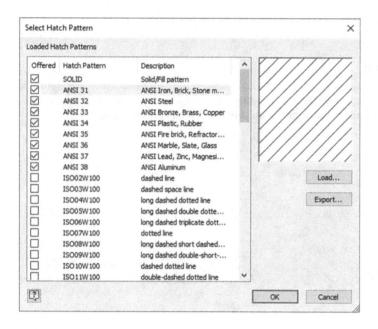

3.   Click **OK**.

Applying Dimensions

Follow these steps:

1.   Click **Annotate ➤ Dimension ➤ Dimension** on the ribbon.

2.   Select the center line on the slot located at the left.

3.   Select the endpoint of the center line of the hole located at the center.

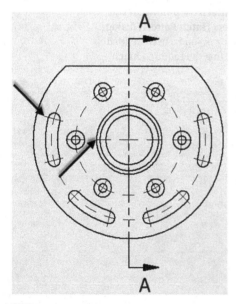

4. Move the pointer toward the left and click.

5. Click **OK**.

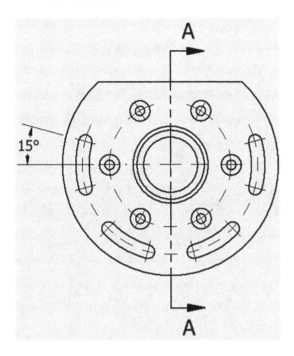

6. Likewise, create another angular dimension, as shown here:

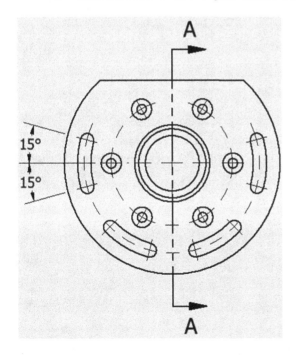

7. Create angular dimensions between the holes and then between slots.

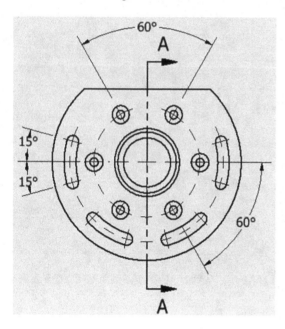

8. Dimension the pitch circle radius of the slots.

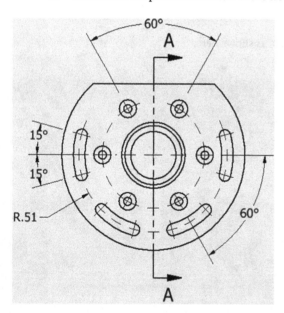

9. With the **Dimension** tool active, select the horizontal line of the front view and the lower quadrant point of the view.

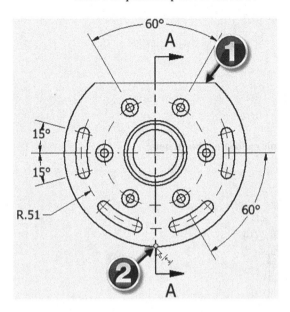

10. Place the dimension on the right side. Click **OK**.

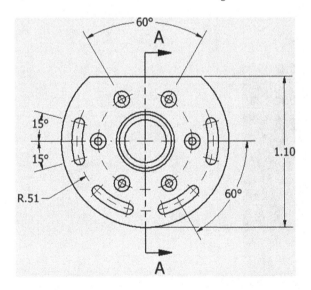

11.　Click **Annotate ➤ Feature Notes ➤ Hole and Thread** on the ribbon.

12.　Select the counterbore hole and place the hole callout, as shown here:

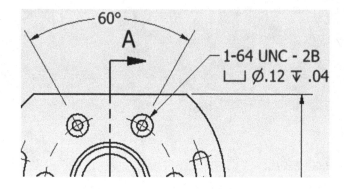

13.　Add a pitch circle radius to counter holes.

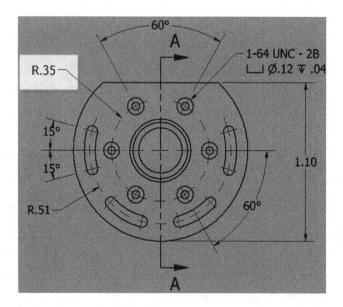

14. Click **Leader Text** on the **Text** panel.

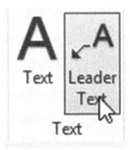

15. Select the slot end, as shown here:

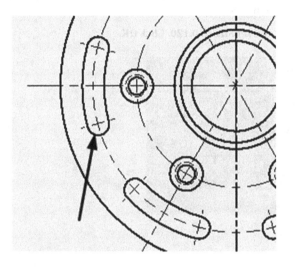

16. Move the cursor away and click.

17. Right-click and select **Continue**; the **Format Text** dialog appears.

18. Enter the text shown here:

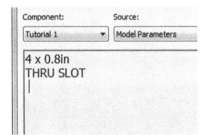

19.   Click **OK**. Press the Esc key.

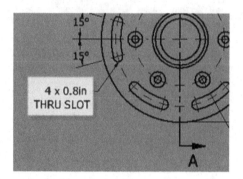

20.   Double-click the section label below the section view.

21.   In the **Format Text** dialog, select all the text and set **Size** to **0.120**. Click **OK**.

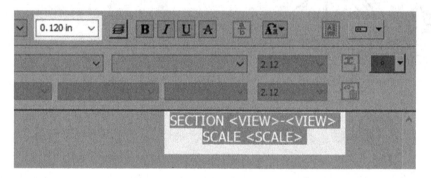

22.   Drag and place the section label on the top.

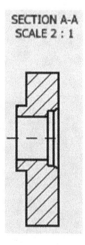

23. Click **Dimension** on the **Dimension** panel.

24. Select the lines, as shown here:

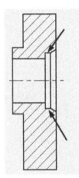

25. Move the pointer toward the right and click to place the dimension.

26. In the dialog, click the **Precision and Tolerance** tab.

27. Set **Tolerance Method** to "**Limits/Fits - Show tolerance.**"

28. Select **Hole ➤ H7**.

29. Set **Primary Unit** value to **3.123**.

30. Set **Primary Tolerance** value to **3.123**.

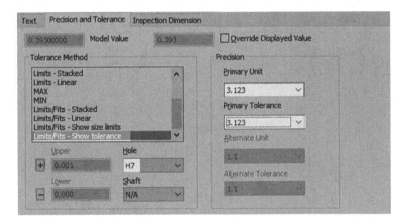

31.   Click **OK**.

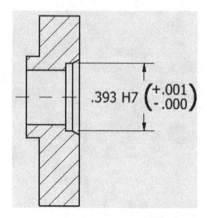

32.   Likewise, apply the other dimensions, as shown here. You can also use the
       Retrieve Dimensions tool to create the dimensions.

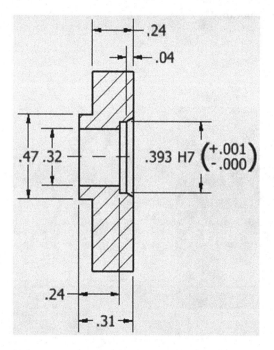

Placing the Datum Feature

Follow these steps:

1.   Click **Annotate ➤ Symbols ➤ Datum Feature** on the ribbon.

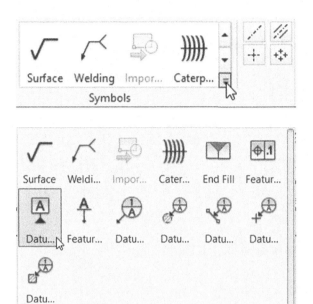

2.   Select the extension line of the dimension, as shown here:

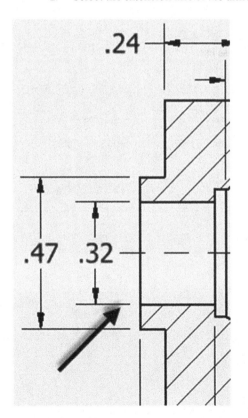

3. Move the cursor downward and click.

4. Move the cursor toward the left and click; the **Format Text** dialog appears. Make sure that **A** is entered in the dialog.

5. Click **OK**.

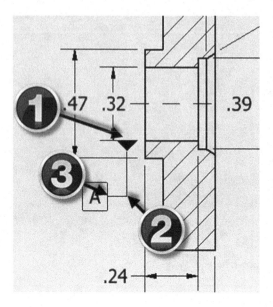

6. Likewise, place a datum feature B, as shown here. Press Esc.

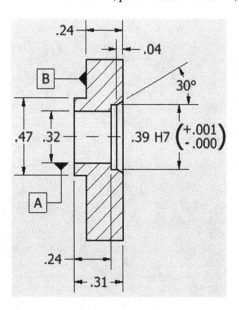

Placing the Feature Control Frame

Follow these steps:

1. Click **Annotate ➤ Symbols ➤ Feature Control Frame** on the ribbon.

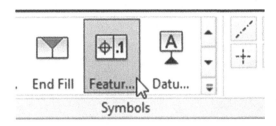

2. Select a point on the line, as shown here.

3. Move the cursor horizontally toward the right and click.

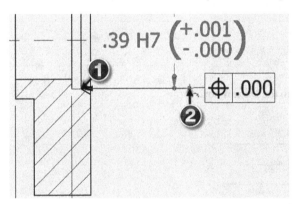

4. Right-click and select **Continue**; the **Feature Control Frame** dialog appears.

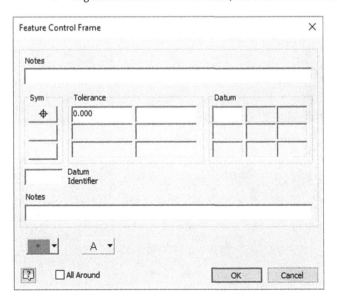

5. In the dialog, click the **Sym** button and select **Circular Run-out**.

6. Enter **0.001** in the **Tolerance** box and **A** in the **Datum** box.

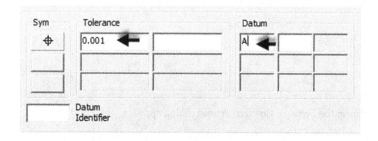

7. Click **OK**.

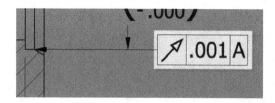

8. Right-click and select **Cancel**.

Placing the Surface Texture Symbols

Follow these steps:

1. Click **Annotate ➤ Symbols ➤ Surface Texture Symbol** on the ribbon.

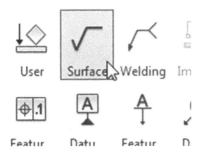

2. Click the inner cylindrical face of the hole, as shown here:

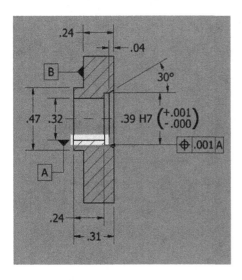

3. Right-click and select **Continue**; the **Surface Texture** dialog appears.

4. Set the **Roughness Average maximum** value to 63.

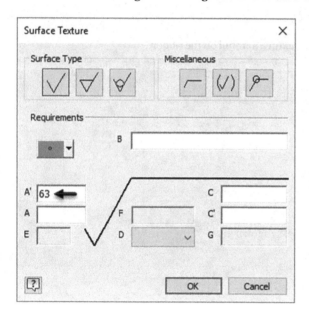

5. Click **OK**.

6. Right-click and select **Cancel**.

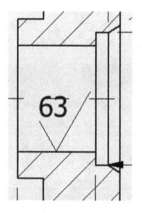

7. Apply the other annotations of the drawing. The final drawing is shown here:

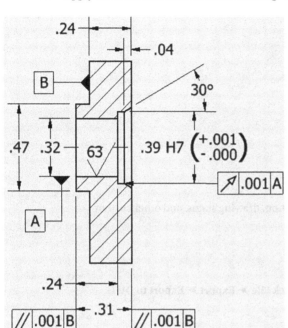

Modifying the Title Block Information

Follow these steps:

1. Right-click the **adapter plate** in the **Browser window**. Select **iProperties** from the shortcut menu.

2.   Click the **Summary** tab and enter the information, as shown here:

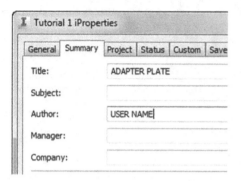

You can also update the project information, drawing status, and other custom information on the respective tabs.

3.   Click **OK**.

4.   Save the file.

5.   To export the file to AutoCAD format, click **File ➤ Export ➤ Export to DWG**.

6.   Click **Save**.

7.   Close the file.

■ ■ ■

CHAPTER 9

Model-Based Dimensioning

Geometric Dimensioning and Tolerancing

During the manufacturing process, the accuracy of a part is an important factor. However, it is impossible to manufacture a part with the exact dimensions. Therefore, while applying dimensions to a drawing, you need to provide some dimensional tolerances that lie within acceptable limits. The following figure shows an example of dimensional tolerances applied to the drawing:

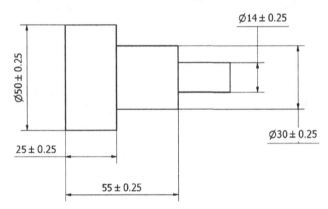

The dimensional tolerances help you to manufacture the component within a specific size range. However, the dimensional tolerances are not sufficient for manufacturing a component. You must give tolerance values to its shape, orientation, and position as well. The following figure shows a note, which is used to explain the tolerance value given to the shape of the object:

Note: The vertical face should not taper over 0.08 from the horizontal face

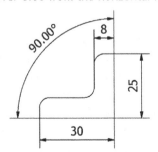

© T. Kishore 2017

T. Kishore, *Learn Autodesk Inventor 2018 Basics*, https://doi.org/10.1007/978-1-4842-3225-5_9

Providing a note in a drawing may be confusing. To avoid this, you can use Geometric Dimensioning and Tolerancing (GD&T) symbols to specify the tolerance values to shape, orientation, and position of a component. The following figure shows the same example represented with GD&T symbols. In this figure, the vertical face to which the tolerance frame is connected must be within two parallel planes 0.08 apart and perpendicular to the datum reference (horizontal plane).

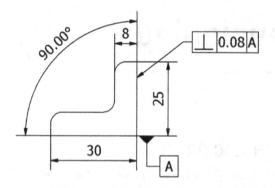

Providing GD&T in 2D drawings is a common and well-known method. You can provide GD&T information to 3D models as well. The tools available on the **Annotate** tab of the ribbon help you to add GD&T information to 3D models based on universal standards such as ASME Y14.41-2003 and ISO 16792: 2006. However, you can add GD&T information based on your custom standard as well.

In this chapter, you will learn to use **Annotate** tools to add GD&T information to the part models. There are many ways to add GD&T information and full-define the parts and assemblies. There are few methods explained in this chapter, but you need to use a method that is most suitable to your design.

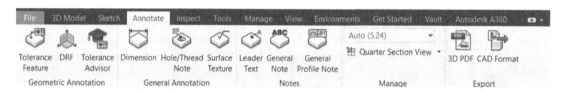

Tutorial 1

This tutorial teaches you to extract dimensions.

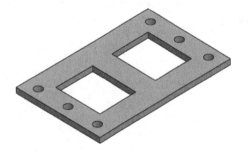

1. Download the model-based dimensioning part files from the companion web site (www.apress.com) and open the Tutorial 1 file.

2. On the ribbon, click **Tools ➤ Options ➤ Document Settings** to open the **Document Settings** dialog.

3. Click the **Standard** tab and select **ASME** from the **Active Standard** drop-down.

4. Click **OK**.

5. In the Browser window, expand the **View** node and then double-click the Isometric view.

6. Right-click the Isometric view and then select **Annotation Scale ➤ Auto**.

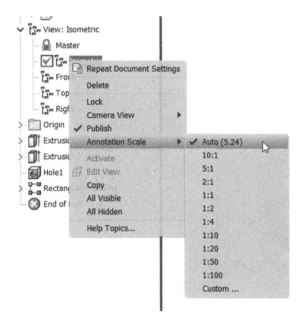

You can also change the **annotation scale** from the **Annotation Scale** drop-down available on the **Manage** panel of the **Annotate** ribbon tab.

Adding Tolerances to the Model Dimensions

Follow these steps:

1. In the Browser window, right-click the Extrusion 1 feature and then select **Show Dimensions**.

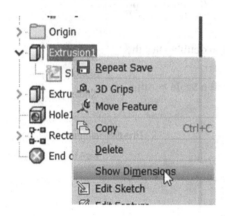

The dimensions of the feature are displayed.

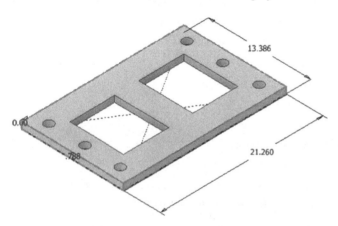

2. Double-click the 13.386 dimension.

3. In the **Edit Dimension** dialog, click the arrow button pointing toward the right and then select **Tolerance**.

4. In the **Tolerance** dialog, select **Type ➤ Symmetric**.

5. Type **0.002** in the **Upper** limit box.

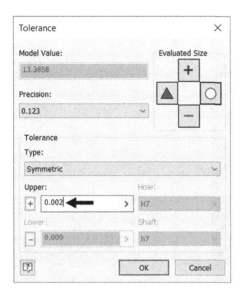

6. Click **OK** in the dialog.

7. Click the green check in the **Edit Dimension** box.

8. Likewise, add tolerances to the remaining dimensions, as shown here:

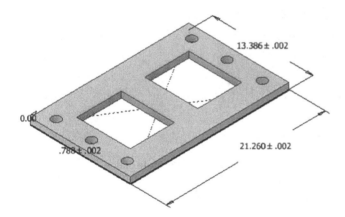

9. Right-click the Extrusion2 feature and then select **Edit Sketch**.

10. Add tolerances to the dimensions, as shown here:

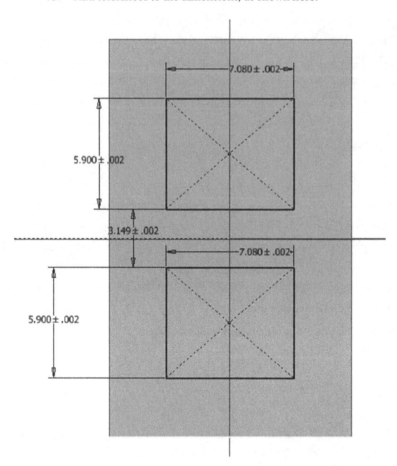

11. Click **Finish Sketch** on the ribbon.

12. Double-click the **Hole** feature in the Browser window to display the **Hole** dialog.

13. In the **Hole** dialog, click the arrow pointing toward the right and then select **Tolerance**.

14. In the **Tolerance** dialog, select **Type ➤ Symmetric**.

15. Type **.002** in the **Upper** limit box and then click **OK**.

16. Zoom to the hole feature and then click the location dimension, as shown.

17. Click the arrow pointing toward the right and then select **Tolerance**.

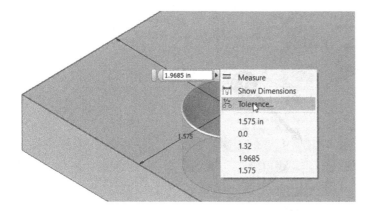

18. In the **Tolerance** dialog, select **Type ➤ Symmetric**.

19. Type **0.002** in the **Upper** limit box and then click **OK**.

20. Likewise, add .002 tolerance to the remaining location dimension. Click **OK** in the **Tolerance** dialog.

21. Click **OK** in the **Hole** dialog.

22. In the Browser window, right-click Rectangular Pattern1 and then select **Show Dimensions**.

23. Add tolerances to the dimensions, as shown here:

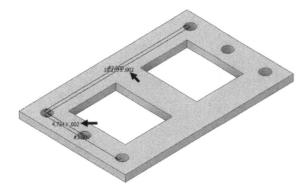

Extracting the Model Dimensions

Follow these steps:

1. In the Browser window, right-click the Extrusion1 feature and then select **Show Dimensions**.

2. Select the 21.260 dimension.

3. Right-click and then select **Promote**.

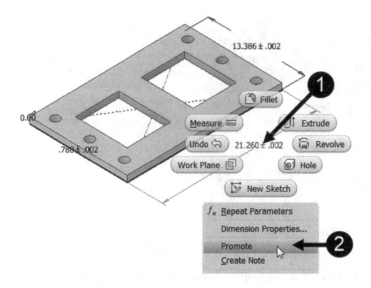

The dimension is promoted as a 3D annotation.

4. Likewise, promote the other two dimensions of the Extrusion1 feature.

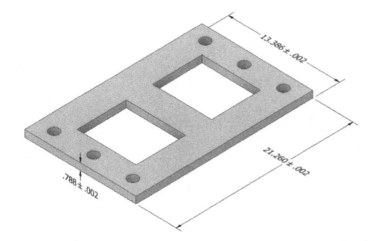

5. Likewise, extract dimensions from the Extrusion2, Hole, and Rectangular Pattern features.

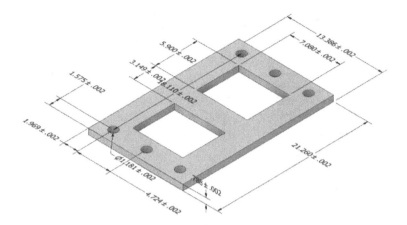

Adding Tolerance Feature

Follow these steps:

1. On the ribbon, click **Annotate ➤ Geometric Annotation ➤ Tolerance Feature**

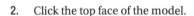

2. Click the top face of the model.

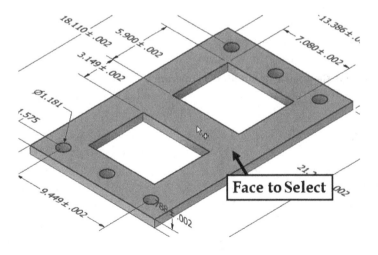

Face to Select

3. Select the **Planar Surface** option from the mini-toolbar.

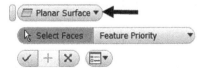

4. Click **OK** on the mini-toolbar.

5. Move the pointer and click to place the tolerance feature.

6. In the tolerance feature, click the tolerance value and then type **.002** in the Tolerance box.

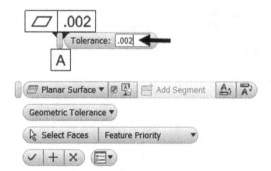

7. Click **OK** on the mini-toolbar.

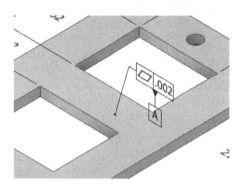

8. On the ribbon, click **Annotate ➤ Geometric Annotation ➤ Tolerance Feature**  .

9. Click the left face of the model.

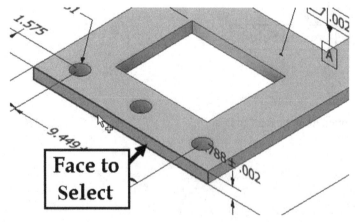

10. Click **OK** on the mini-toolbar.

11. Move the pointer and click to place the tolerance feature.

12. In the tolerance feature, click the tolerance value and type **.002** in the Tolerance box.

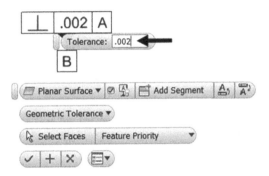

13. Click **OK** on the mini-toolbar.

14. Likewise, create another tolerance feature, as shown here:

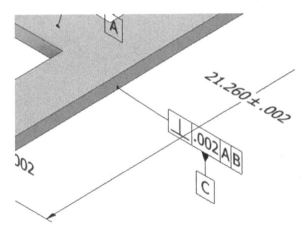

15. Select the hole annotation and press Delete on your keyboard.

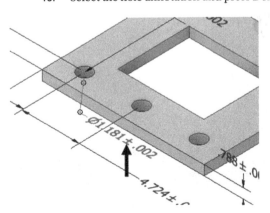

16. On the ribbon, click **Annotate** ➤ **Geometric Annotation** ➤ **Tolerance Feature** .

17. Select the Hole feature from the model.

18. Select **Simple Hole Parallel Axis Pattern** from the mini-toolbar.

19. Click **OK** on the mini-toolbar.

20. Right-click and **Select Annotation Plane** [Shift].

21. Select the top face of the model.

22. Click to place the hole annotation.

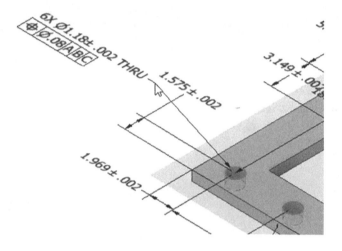

23. Click the tolerance value and then type **.002** in the **Tolerance** box.

24. Select **Maximum Material Condition** from the drop-down available next to the **Tolerance** box.

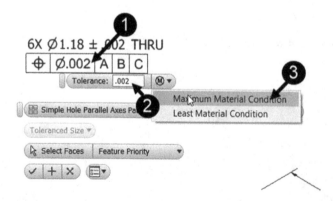

25.   Click **OK** on the mini-toolbar.

The annotations and tolerance features are listed in the Browser window.

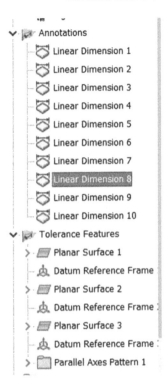

26.   Save and close the part file.

Index

P, Q

R

S

T, U, V, W, X, Y

Z

Get the eBook for only $5!

Why limit yourself?

With most of our titles available in both PDF and ePUB format, you can access your content wherever and however you wish—on your PC, phone, tablet, or reader.

Since you've purchased this print book, we are happy to offer you the eBook for just $5.

To learn more, go to http://www.apress.com/companion or contact support@apress.com.

Apress®

Get the eBook for only $5!

Why limit yourself?

With most of our titles available in both PDF and ePUB format, you can access your content wherever and however you wish—on your PC, phone, tablet, or reader.

Since you've purchased this print book, we are happy to offer you the eBook for just $5.

To learn more, go to http://www.apress.com/companion or contact support@apress.com.

Apress®